21 世纪全国高职高专计算机系列实用规划教材

计算机专业英语教程(第 2 版)

主　编　李　莉　李秀华
副主编　樊晋宁　康丽军
参　编　黄晓霞　李　茜
　　　　彭卫华

内 容 简 介

本书内容包括计算机基础知识、计算机网络、Internet 应用、数据库技术、编程语言、信息安全、图像处理、嵌入式系统等 8 个方面，涉及计算机应用的常见领域和最新技术以及与职业岗位有关的实用知识。所有材料选自网络文献及原版精品教材，每篇课文后有典型例句的语法分析，以及与课文内容有关的练习题和阅读材料。

本书针对高等职业院校培养高素质技能型专门人才的需要编写，从实用出发，扩大读者计算机应用领域的词汇量和知识面，提高计算机应用相关资料的阅读能力。本书难度适中，图文并茂，可以作为高职院校计算机专业和信息技术专业的教材，也可以作为计算机工程技术人员提高英语能力的自学用书。

图书在版编目(CIP)数据

计算机专业英语教程/李莉，李秀华主编.—2 版. —北京：北京大学出版社，2010.1
(21 世纪全国高职高专计算机系列实用规划教材)
ISBN 978-7-301-16046-6

Ⅰ.计… Ⅱ.①李…②李… Ⅲ.电子计算机－英语－高等学校：技术学校－教材 Ⅳ.H31

中国版本图书馆 CIP 数据核字(2009)第 197689 号

书　　　名：计算机专业英语教程(第 2 版)
著作责任者：李　莉　李秀华　主编
责 任 编 辑：刘国明
标 准 书 号：ISBN 978-7-301-16046-6/TP·1062
出　版　者：北京大学出版社
地　　　址：北京市海淀区成府路 205 号　100871
网　　　址：http://www.pup.cn　http://www.pup6.com
电　　　话：邮购部 62752015　发行部 62750672　编辑部 62750667　出版部 62754962
电 子 邮 箱：pup_6@163.com
印　刷　者：北京飞达印刷有限责任公司
发　行　者：北京大学出版社
经　销　者：新华书店
　　　　　　787mm×1092mm　16 开本　16.75 印张　378 千字
　　　　　　2005 年 9 月第 1 版　2010 年 1 月第 2 版　2016 年 7 月第 4 次印刷
定　　　价：26.00 元

未经许可，不得以任何方式复制或抄袭本书之部分或全部内容。
版权所有　侵权必究　　举报电话：010-62752024
　　　　　　　　　　　电子邮箱：fd@pup.pku.edu.cn

21世纪全国高职高专计算机系列实用规划教材
专家编审委员会

主　任　　刘瑞挺

副主任　　(按拼音顺序排名)

　　　　　　陈玉国　　崔锁镇　　高文志　　韩希义

　　　　　　黄晓敏　　魏　峥　　谢一风　　张文学

委　员　　(按拼音顺序排名)

　　　　　　安志远　丁亚明　杜兆将　高爱国　高春玲　郭鲜凤
　　　　　　韩最蛟　郝金镇　黄贻彬　季昌武　姜　力　李晓桓
　　　　　　连卫民　刘德军　刘德仁　刘辉珞　栾昌海　罗　毅
　　　　　　慕东周　彭　勇　齐彦力　沈凤池　陶　洪　王春红
　　　　　　闻红军　武凤翔　武俊生　徐　红　徐洪祥　徐受容
　　　　　　许文宪　严仲兴　杨　武　易永红　于巧娥　袁体芳
　　　　　　张　昕　赵　敬　赵润林　周朋红　訾　波　周　奇

信息技术的职业化教育

(代丛书序)

刘瑞挺/文

北京大学出版社第六事业部组编了一套《21世纪全国高职高专计算机系列实用规划教材》。为此，制订了详细的编写目的、丛书特色、内容要求和风格规范。在内容上强调面向职业、项目驱动、注重实例、培养能力；在风格上力求文字精练、图表丰富、脉络清晰、版式明快。

一、组编过程

2004年10月，第六事业部开始策划这套丛书，分派编辑深入各地职业院校，了解教学第一线的情况，物色经验丰富的作者。2005年1月15日在济南召开了"北大出版社高职高专计算机规划教材研讨会"。来自13个省、41所院校的70多位教师汇聚一堂，共同商讨未来高职高专计算机教材建设的思路和方法，并对规划教材进行了讨论与分工。2005年6月13日在苏州又召开了"高职高专计算机教材大纲和初稿审定会"。编审委员会委员和45个选题的主、参编，共52位教师参加了会议。审稿会分为公共基础课、计算机软件技术专业、计算机网络技术专业、计算机应用技术专业4个小组对稿件逐一进行审核。力争编写出一套高质量的、符合职业教育特点的精品教材。

二、知识结构

职业生涯的成功与人们的知识结构有关。以著名侦探福尔摩斯为例，作家柯南道尔在"血字的研究"中，对其知识结构描述如下：

- ◆ 文学知识——无；
- ◆ 哲学知识——无；
- ◆ 政治学知识——浅薄；
- ◆ 植物学知识——不全面。对于药物制剂和鸦片却知之甚详。对毒剂有一般了解，而对于实用园艺却一无所知；
- ◆ 化学知识——精深；
- ◆ 地质学知识——偏于应用，但也有限。他一眼就能分辨出不同的土质。根据裤子上泥点的颜色和坚实程度就能说明是在伦敦什么地方溅上的；
- ◆ 解剖学知识——准确，却不系统；
- ◆ 惊险小说知识——很渊博。似乎对近一个世纪发生的一切恐怖事件都深知底细；
- ◆ 法律知识——熟悉英国法律，并能充分实用；
- ◆ 其他——提琴拉得很好，精于拳术、剑术。

事实上，我国唐朝名臣狄仁杰，大宋提刑官宋慈，都有类似的知识结构。审视我们自己，每人的知识结构都是按自己的职业而建构的。因此，我们必须面向职场需要来设计教材。

三、职业门类

我国的职业门类分为 18 个大类：农林牧渔、交通运输、生化与制药、地矿与测绘、材料与能源、土建水利、制造、电气信息、环保与安全、轻纺与食品、财经、医药卫生、旅游、公共事业、文化教育、艺术设计传媒、公安、法律。

每个职业大类又分为二级类，例如电气信息大类又分为 5 个二级类：计算机、电子信息、通信、智能控制、电气技术。因此，18 个大类共有 75 个二级类。

在二级类的下面，又有不同的专业。75 个二级类共有 590 种专业。俗话说："三百六十行，行行出状元"，现代职业仍在不断涌现。

四、IT 能力领域

通常信息技术分为 11 个能力领域：规划的能力、分析与设计 IT 解决方案的能力、构建 IT 方案的能力、测试 IT 方案的能力、实施 IT 方案的能力、支持 IT 方案的能力、应用 IT 方案的能力、团队合作能力、文档编写能力、项目管理能力以及其他能力。

每个能力领域下面又包含若干个能力单元，11 个能力领域共有 328 个能力单元。例如，应用 IT 方案能力领域就包括 12 个能力单元。它们是操作计算机硬件的能力、操作计算软件包的能力、维护设备与耗材的能力、使用计算软件包设计机构文档的能力、集成商务计算软件包的能力、操作文字处理软件的能力、操作电子表格应用软件的能力、操作数据库应用软件的能力、连接到互联网的能力、制作多媒体网页的能力、应用基本的计算机技术处理数据的能力、使用特定的企业系统以满足用户需求的能力。

显然，不同的职业对 IT 能力有不同的要求。

五、规划梦想

于是我们建立了一个职业门类与信息技术的平面图，以职业门类为横坐标、以信息技术为纵坐标。每个点都是一个函数，即 IT(Professional)，而不是 IT+Professional 单纯的相加。针对不同的职业，编写它所需要的信息技术教材，这是我们永恒的主题。

这样组合起来，就会有 IT((328)*(Pro(590)))，这将是一个非常庞大的数字。组织这么多的特色教材，真的只能是一个梦想，而且过犹不及。能做到 IT((11)*(Pro(75)))也就很不容易了。

因此，我们既要在宏观上把握职业门类的大而全，也要在微观上选择信息技术的少而精。

六、精选内容

在计算机科学中，有一个统计规律，称为 90/10 局部性原理(Locality Rule)：即程序执行的 90%代码，只用了 10%的指令。这就是说，频繁使用的指令只有 10%，它们足以完成 90%的日常任务。

事实上，我们经常使用的语言文字也只有总量的 10%，却可以完成 90%的交流任务。同理，我们只要掌握了信息技术中 10%频繁使用的内容，就能处理 90%的职业化任务。

有人把它改为 80/20 局部性原理，似乎适应的范围更广些。这个规律为编写符合职业教育需要的精品教材指明了方向：坚持少而精，反对多而杂。

七、职业本领

以计算机为核心、贴近职场需要的信息技术已经成为大多数人就业的关键本领。职业教育的目标之一就是培养学生过硬的 IT 从业本领，而且这个本领必须上升到职业化的高度。

职场需要的信息技术不仅是会使用键盘、录入汉字，而且还要提高效率、改善质量、降低成本。例如，两位学生都会用 Office 软件，但他们的工作效率、完成质量、消耗成本可能有天壤之别。领导喜欢谁？这是不言而喻的。因此，除了道德品质、工作态度外，必须通过严格的行业规范和个人行为规范，进行职业化训练才能养成正确的职业习惯。

我们肩负着艰巨的历史使命。我国人口众多，劳动力供大于求的矛盾将长期存在。发展和改革职业教育，是我国全面建设小康社会进程中一项艰巨而光荣的任务，关系到千家万户人民群众的切身利益。职业教育和高技能人才在社会主义现代化建设中有特殊的作用。我们一定要兢兢业业、不辱使命，把这套高职高专教材编写好，为我国职业教育的发展贡献一份力量。

刘瑞挺教授 曾任中国计算机学会教育培训委员会副主任、教育部理科计算机科学教学指导委员会委员、全国计算机等级考试委员会委员。目前担任的社会职务有：全国高等院校计算机基础教育研究会副会长、全国计算机应用技术证书考试委员会副主任、北京市计算机教育培训中心副理事长。

本系列教材编写目的和教学服务

本系列教材在遍布全国的各位编写老师的共同辛勤努力下,在编委会主任刘瑞挺教授和其他编审委员会成员的指导下,在北京大学出版社第六事业部的各位编辑刻苦努力下,本系列教材终于与广大师生们见面了。

教材编写目的

近几年来,职业技术教育事业得以蓬勃的发展,全国各地的高等职业院校以及高等专科学校无论是从招生人数还是学校的软、硬件设施上都达到了相当规模。随着我国经济的高速发展,尽快提高职业技术教育的水平显得越来越重要。教育部提出:职业教育就是就业教育,也就是说教学要直接面对就业,强调实践。不但要介绍技术,更要介绍具体应用,注重技术与应用的结合。本套教材的主要编写思想如下。

1. 与发达国家相比,我国职业技术教育教材的发展比较缓慢并且滞后,远远跟不上职业技术教育发展的需求。我们常常提倡职业教育的实用性,但在课堂教学中仍然使用理论性和技术性教材进行职业实践教学。针对这种现状,急需推出一系列切合当前教育改革需要的高质量的优秀职业技术实训型教材。

2. 本套教材总结了目前优秀计算机职业教育专家的教学思想与经验,与广大职业教育一线老师共同探讨,最终落实到本套教材中,开发出一套适合于我国职业教育教学目标和教学要求的教材,它是一套能切实提高学生专业动手实践能力和职业技术素质的教材。

3. 社会对学生的职业能力的要求不断提高,从而催化出了许多新型的课程结构和教学模式。新型教学模式是必须以工作为基础的模仿学习,它是将学生置于一种逼真的模拟环境中,呈现给学生的是具有挑战性、真实性和复杂性的问题,使学生得到较真实的锻炼。

4. 教材的结构必须按照职业能力的要求创建并组织实施新的教学模式。教学以专项能力的培养展开,以综合能力的形成为目标。能力的培养既是教学目标,又是评估的依据和标准。

5. 本套的重点是先让学生实践,从实践中领悟、总结理论,然后再学习必要的理论,用理论指导实践。从这一个循环的教学过程中,学生的职业能力将得到极大的提高。

教学服务

1. 提供电子教案

本系列教材绝大多数都是教程与实训二合一,每一本书都有配套的电子教案,以降低任课老师的备课强度,此课件可以在我们网站上随时下载。

2. 提供教学资源下载

本系列教材中涉及到的实例(习题)的原始图片和其他素材或者是源代码、原始数据等文件,都可以在我们网站上下载。

3. 提供多媒体课件和教师培训

针对某些重点课程,我们配套有相应的多媒体课件。对大批量使用本套教材的学校,我们会免费提供多媒体课件,另外还将免费提供教师培训名额,组织使用本套教材的教师进行相应的培训。

北京大学出版社第六事业部(http://www.pup6.com)

第 2 版前言

为了适应高等职业院校培养高素质技能型专门人才的需要，高职高专教材应该反映工程技术领域的进步和更新。鉴于计算机技术及应用发展迅速，我们于 2008 年开始编写《计算机专业英语教程》(第 2 版)，主要进行了如下工作：

- 对第 1 版中的错误进行了修改。
- 对内容作了调整更新，去掉了陈旧过时的部分，补充了有关 Flash Memory、Windows Vista、Embedded System 等新技术的相关知识以及与职业岗位有关的实用内容，如 Software Engineer、Database Administrator、Guidelines for Writing Software Documentation 等，使得本书更适应高职高专的教学需要。
- 将第 1 版中篇幅较长的章节做了修改，使得每一节的材料难度和篇幅都较为适中，并且每一节的内容紧扣一个主题。
- 根据第 1 版在使用过程中的反馈意见，增加了阅读材料的参考译文。
- 增加了插图，使得本书图文并茂，内容更加生动，更易于理解。

参加《计算机专业英语教程》(第 2 版)编写工作的有山东电力高等专科学校的李莉、李秀华，太原城市职业技术学院的樊晋宁，太原大学的康丽军，山东聊城职业技术学院的黄晓霞，山东商业职业技术学院的李茜，湖南对外经济贸易职业学院的彭卫华。本书由李莉提出编写计划与方案，并编写了第 5 章、第 7 章的正文部分以及 6.1 节；李秀华编写第 8 章以及第 1、2、3、4 章的阅读材料部分；樊晋宁编写了第 1、2 章的正文部分；康丽军编写了第 3 章的正文部分；黄晓霞编写了第 6 章的 6.2、6.3、6.4 节；李茜编写了第 4 章的正文部分；彭卫华编写了第 5、7 章的阅读材料部分。最后全书统稿由李莉担任；李毅协助完成了许多文字录入和校对工作。

在编写过程中，刘瑞挺教授和多所院校的老师对本书的编写提出了非常好的建议，在此作者表示衷心的感谢！

由于编者水平所限，书中内容可能会有错误和疏漏之处，敬请读者不吝赐教。编者联系方式：lily@sdu.edu.cn。

编　者
2009 年 12 月

目 录

Chapter 1　Computer Fundamentals 1
- 1.1　Four Kinds of Computers 1
  - 1.1.1　Reading Material 5
  - 1.1.2　正文参考译文 7
  - 1.1.3　阅读材料参考译文 8
- 1.2　Computer Hardware 9
  - 1.2.1　Reading Material 15
  - 1.2.2　正文参考译文 16
  - 1.2.3　阅读材料参考译文 18
- 1.3　System Software 18
  - 1.3.1　Reading Material 23
  - 1.3.2　正文参考译文 24
  - 1.3.3　阅读材料参考译文 26
- 1.4　Application Software 26
  - 1.4.1　Reading Material 31
  - 1.4.2　正文参考译文 33
  - 1.4.3　阅读材料参考译文 34

Chapter 2　Computer Network 36
- 2.1　Introduction to Computer Network 36
  - 2.1.1　Reading Material 40
  - 2.1.2　正文参考译文 42
  - 2.1.3　阅读材料参考译文 44
- 2.2　Data Communications Channels 44
  - 2.2.1　Reading Material 49
  - 2.2.2　正文参考译文 51
  - 2.2.3　阅读材料参考译文 52
- 2.3　Main Factors Affecting Data Transmission 53
  - 2.3.1　Reading Material 57
  - 2.3.2　正文参考译文 59
  - 2.3.3　阅读材料参考译文 60
- 2.4　Network Architecture 61
  - 2.4.1　Reading Material 66
  - 2.4.2　正文参考译文 67
  - 2.4.3　阅读材料参考译文 69

Chapter 3　Internet Applications 71
- 3.1　Browsers and E-mails 71
  - 3.1.1　Reading Material 75
  - 3.1.2　正文参考译文 76
  - 3.1.3　阅读材料参考译文 77
- 3.2　Search Tools 78
  - 3.2.1　Reading Material 82
  - 3.2.2　正文参考译文 84
  - 3.2.3　阅读材料参考译文 85
- 3.3　Definitions and Content of the Electronic Commerce 86
  - 3.3.1　Reading Material 91
  - 3.3.2　正文参考译文 92
  - 3.3.3　阅读材料参考译文 94
- 3.4　Value Chains in E-commerce 94
  - 3.4.1　Reading Material 99
  - 3.4.2　正文参考译文 101
  - 3.4.3　阅读材料参考译文 102

Chapter 4　Database Fundamentals 104
- 4.1　Introduction to DBMS 104
  - 4.1.1　Reading Material 108
  - 4.1.2　正文参考译文 111
  - 4.1.3　阅读材料参考译文 112
- 4.2　Structure of the Relational Database I 113
  - 4.2.1　Reading Material 117
  - 4.2.2　正文参考译文 119
  - 4.2.3　阅读材料参考译文 120
- 4.3　Structure of the Relational Database II 121

	4.3.1	Reading Material 125
	4.3.2	正文参考译文 126
	4.3.3	阅读材料参考译文 127
4.4	Structured Query Language 128
	4.4.1	Reading Material 132
	4.4.2	正文参考译文 134
	4.4.3	阅读材料参考译文 135

Chapter 5　Programming Language 138

5.1	Algorithms and Flowcharts 138
	5.1.1	Reading Material 141
	5.1.2	正文参考译文 143
	5.1.3	阅读材料参考译文 144
5.2	Introduction of Programming Languages 145
	5.2.1	Reading Material 149
	5.2.2	正文参考译文 150
	5.2.3	阅读材料参考译文 151
5.3	Object-Oriented Programming 152
	5.3.1	Reading Material 155
	5.3.2	正文参考译文 157
	5.3.3	阅读材料参考译文 158
5.4	Program Debugging and Program Maintenance 159
	5.4.1	Reading Material 163
	5.4.2	正文参考译文 166
	5.4.3	阅读材料参考译文 167

Chapter 6　Information Security 169

6.1	Concept of Information Security 169
	6.1.1	Reading Material 172
	6.1.2	正文参考译文 174
	6.1.3	阅读材料参考译文 175
6.2	Computer Viruses 175
	6.2.1	Reading Material 179
	6.2.2	正文参考译文 181
	6.2.3	阅读材料参考译文 182
6.3	Internet Security 183
	6.3.1	Reading Material 186

	6.3.2	正文参考译文 187
	6.3.3	阅读材料参考译文 189
6.4	Secure Networks and Policies 189
	6.4.1	Reading Material 192
	6.4.2	正文参考译文 193
	6.4.3	阅读材料参考译文 194

Chapter 7　Image Processing 195

7.1	Concepts of Graphic and Image 195
	7.1.1	Reading Material 198
	7.1.2	正文参考译文 200
	7.1.3	阅读材料参考译文 200
7.2	Introduction to Digital Image Processing 201
	7.2.1	Reading Material 206
	7.2.2	正文参考译文 207
	7.2.3	阅读材料参考译文 208
7.3	Image Compression 209
	7.3.1	Reading Material 212
	7.3.2	正文参考译文 213
	7.3.3	阅读材料参考译文 214
7.4	Application of Digital Image Processing 215
	7.4.1	Reading Material 219
	7.4.2	正文参考译文 221
	7.4.3	阅读材料参考译文 222

Chapter 8　Embedded Systems 224

8.1	Come to Study Embedded Systems 224
	8.1.1	Reading Material 227
	8.1.2	正文参考译文 228
	8.1.3	阅读材料参考译文 229
8.2	Characteristics of Embedded Systems 230
	8.2.1	Reading Material 232
	8.2.2	正文参考译文 234
	8.2.3	阅读材料参考译文 234
8.3	System-level Requirements 235
	8.3.1	Reading Material 239

		8.3.2 正文参考译文 241

    8.3.3 阅读材料参考译文 242

8.4 Application of Embedded System 243

    8.4.1 Reading Material 246

    8.4.2 正文参考译文 249

    8.4.3 阅读材料参考译文..................249

参考文献 251

参考文章 251

Chapter 1　Computer Fundamentals

1.1　Four Kinds of Computers

Computers are electronic devices that can follow instructions to accept input, process that input, and produce information. There are four types of computers: microcomputers, minicomputers, mainframe computers, and supercomputers.

Microcomputers, also known as personal computers, are small computers that can fit on a desktop. Portable microcomputers can fit in a briefcase or even in the palm of your hand. Microcomputers are used in homes, schools, and industry. Today nearly every field uses microcomputers.

One type of microcomputer that is rapidly growing in popularity is the portable computer, which can be easily carried around. There are four categories of portable computers.

Laptops: laptops, which weigh between 10 and 16 pounds, may be AC-powered, battery-powered, or both. The AC-powered laptop weighs 12 to 16 pounds. The battery-powered laptop weighs 10 to 15 pounds, batteries included, and can be carried on a shoulder strap. Figure 1.1 shows an example of a laptop.

Figure 1.1　A Modern Mid-range HP Laptop

Notebook PCs: notebook personal computers weigh between 5 and 10 pounds and can fit into most briefcases. It is especially valuable in locations where electrical connections are not available. Notebook computers are the most popular portable computers today.

Subnotebooks: subnotebooks are for frequent flyers and life-on-the-road types. [1]Subnotebooks users give up a full-size display screen and keyboard in exchange for less weight. Weighting between 2 and 6 pounds, these computers fit easily into a briefcase.

Personal Digital Assistants: much smaller than even the subnotebooks. Personal Digital Assistants (PDAs) weigh from 1 to 4 pounds. The typical PDA combines pen input, writing recognition, personal organizational tools, and communication capabilities in a very small package. Figure 1.2 shows an example of a PDA.

Minicomputers, also known as midrange computers, are desk-sized machines. [2]They fall into between microcomputers and mainframes in their processing speeds and data-storing capacities. Medium-size companies or departments of large companies typically use them for specific purposes. For example, they might use them to do research or to monitor a particular manufacturing process. Smaller-size companies typically use microcomputers for their general data processing needs, such as accounting.

Figure 1.2　A PDA

Mainframe computers are larger computers occupying specially wired, air-conditioned rooms and capable of great processing speeds and data storage. They are used by large organizations—business, banks, universities, government agencies—to handle millions of transactions. For example, insurance companies use mainframes to process information about millions of policyholders.

Supercomputers are special, high-capacity computers used by very large organizations principally for research purposes. Among their uses are oil exploration and worldwide weather forecasting. An example of a supercomputer is shown in Figure 1.3.

Figure 1.3　Mississippi State system administrators inspect the new Raptor supercomputer in the university's High Performance Computing Collaboratory

In general, a computer's type is determined by the following seven factors:

The type of CPU. Microcomputers use microprocessors. The larger computers tend to use CPUs made up of separate, high-speed, sophisticated components.

The amount of main memory the CPU can use. A computer equipped with a large amount of main memory can support more sophisticated programs and can even hold several different programs in memory at the same time.

The capacity of the storage devices. The larger computer systems tend to be equipped with higher capacity storage devices.

The speed of the output devices. [3]The speed of microcomputer output devices tends to be rated in terms of the number of characters per second (cps) that can be printed——usually in tens and hundreds of cps. Larger computers' output devices are faster and are usually rated at speeds of hundreds or thousands of lines that can be printed per minute.

The processing speed in millions of instructions per second (MIPS). The term instruction is used here to describe a basic task the software asks the computer to perform while also identifying the data to be affected. The processing speed of the smaller computers ranges from 7 to 40 **MIPS**. The speed of large computers can be 30 to 150 **MIPS** or more, and supercomputers can process more than 200 **MIPS**. In other words, a mainframe computer can process your data a great deal faster than a microcomputer can.

The number of users that can access the computer at one time. Most small computers can support only a single user, some can support as many as two or three at a time. Large computers can support hundreds of users simultaneously.

The cost of the computer system. Business systems can cost as little as $500 (for a microcomputer) or as much as $10 million (for a mainframe)——and much more for a supercomputer.

Words

access	v.	访问，存取
accounting	n.	会计学，统计
agency	n.	机构，办事处，代理店
briefcase	n.	公文包，公事包
exploration	n.	勘探，发掘，调查
laptop	n.	便携式电脑
palm	n.	(手)掌，手心，掌状物
policyholder	n.	投保人，保险客户
recognition	n.	认出，识别，认可
simultaneously	adv.	同时地
sophisticated	adj.	精密复杂的
strap	n.	(皮，布)带
typically	adv.	特有地，独特地，典型地

Phrases

fall into	属于
in exchange for	交换，来代替
mainframe computer	大型计算机
portable computer	便携式计算机

Abbreviations

AC(Alternating Current)　　　　　　　　交流电

PDA(Personal Digital Assistant)　　　　　个人数字助理

Notes

[1] 例句：Subnotebooks users give up a full-size display screen and keyboard in exchange for less weight.

分析：in exchange for less weight 是介词短语作目的状语，意思是"为了换取较轻的重量"。

译文：超小型笔记本电脑用户放弃了大尺寸的显示器和键盘，换来了重量的减轻。

[2] 例句：They fall in between microcomputers and mainframes in their processing speeds and data-storing capacities.

分析：句中 They 指的是小型计算机；短语 fall in between 意思是"介于两者之间"，in their processing speeds and data-storing capacities 是介词短语作状语。

译文：小型计算机的处理速度和数据存储能力介于微型机和大型机之间。

[3] 例句：The speed of microcomputer output devices tends to be rated in terms of the number of characters per second (cps) that can be printed——usually in tens and hundreds of cps.

分析：句中 tends to be rated…，意思是"倾向于用……度量"，tend to do sth. 意思是"倾向于/易于做某事"。例句：They tend to do badly in school. 他们通常在学校表现不好。In the end, everything tends to do what you would expect. 最后，所有的事情都会像你期望的那样运转。

译文：微机输出设备的速度倾向于用每秒钟能打印的字符数(cps)予以度量，通常为每秒几十个、几百个字符。

Exercises

Ⅰ. Put "true" or "false" in the brackets for the following statements according to the passage.

1. (　) Computers are electronic devices that can perform tasks automatically.
2. (　) Portable computers can fit in a briefcase or even in the palm of your hand.
3. (　) Portable computers are AC-powered, battery-powered, or both.
4. (　) All portable computers can fit in briefcases.
5. (　) Subnotebooks have full-size display screens and keyboards.
6. (　) The capacity of the storage devices is a main factor that affects the property of computers.
7. (　) Most microcomputers are single-user systems.
8. (　) According to the passage, supercomputers have storage devices with the largest capacity.

9. ( ) The term "instruction" used in the passage only describes a basic task that the software asks the computer to perform.

10. ( ) Ordinary users have chances to come into contact with supercomputers.

II. Fill in the blanks according to the passage.

1. Computers are _____ devices that can follow _____ to accept input, process that input, and produce information.

2. Portable microcomputers can fit in a _____ or even in the _____ of your hand.

3. There are four types of computers: _____, _____, _____, _____.

4. Minicomputers fall into between microcomputers and mainframes in their _____ speeds and data-storing _____.

5. Mainframe computers are used by large _____.

6. Insurance companies use mainframes to process _____ about millions of _____.

7. Supercomputers are special, _____ computers.

8. Laptops may be _____, _____, or both.

9. The larger computer systems tend to be _____ with higher capacity _____.

10. Large computer can support hundreds of users _____.

III. Translate the following words and expressions into Chinese.

1. electronic device
2. personal computer
3. processing speed
4. high-capacity
5. battery-powered
6. portable computer
7. writing recognition
8. storage device
9. supercomputer
10. mainframe

1.1.1 Reading Material

ENIAC

The title of forefather of today's all-electronic digital computers is usually awarded to ENIAC, which stood for Electronic Numerical Integrator and Calculator. ENIAC was built at the University of Pennsylvania between 1943 and 1945 by two professors, John Mauchly and the 24 years old J. Presper Eckert, who got funding from the war department after promising they could build a machine that would replace all the "computers", meaning the women who were employed calculating the firing tables for the army's artillery guns.

ENIAC filled a 20 by 40 foot room, weighed 30 tons, and used more than 18,000 vacuum tubes. ENIAC employed paper card readers obtained from IBM. When operating, the ENIAC was silent but you knew it was on as the 18,000 vacuum tubes each generated waste heat like a light bulb and all this heat (174,000 watts of heat) meant that the computer could only be operated in a specially designed room with its own heavy duty air conditioning system. Only the left half of ENIAC is visible in the first picture in Figure 1.4, the right half was basically a mirror image of what's visible.

Figure 1.4 Two views of ENIAC: the "Electronic Numerical Integrator and Calculator"

To reprogram the ENIAC you had to rearrange the patch cords that you can observe on the left in the prior photo, and the settings of 3000 switches that you can observe on the right. To program a modern computer, you type out a program with statements like:

Circumference = 3.14 * diameter

To perform this computation on ENIAC you had to rearrange a large number of patch cords and then locate three particular knobs on that vast wall of knobs and set them to 3, 1, and 4.

One of the most obvious problems was that the design would require 18,000 vacuum tubes to all work simultaneously. Vacuum tubes were notoriously unreliable. The idea that 18,000 tubes could function together was considered so unlikely that the dominant vacuum tube supplier of the day, RCA, refused to join the project (but did supply tubes in the interest of "wartime cooperation"). Eckert solved the tube reliability problem through extremely careful circuit design. He was so thorough that before he chose the type of wire cabling he would employ in ENIAC he first ran an experiment where he starved lab rats for a few days and then gave them samples of all the available types of cable to determine which they least liked to eat. Here's a look at a small number of the vacuum tubes in ENIAC:

Even with 18,000 vacuum tubes, ENIAC could only hold 20 numbers at a time. ENIAC's basic clock speed was 100,000 cycles per second. Today's home computers employ clock speeds of 1,000,000,000 cycles per second. Built with $500,000 from the U.S. Army, ENIAC's first task was to compute whether or not it was possible to build a hydrogen bomb (the atomic bomb was completed during the war and hence is older than ENIAC). The very first problem run on ENIAC required only 20 seconds and was checked against an answer obtained after forty hours of work with a mechanical calculator. After chewing on half a million punch cards for six weeks, ENIAC did humanity no favor when it declared the hydrogen bomb feasible. This first ENIAC program remains classified even today.

Eckert and Mauchly's next teamed up with the mathematician **John von Neumann** to design EDVAC, which pioneered the **stored program**.

Words

artillery	n.	炮
bulb	n.	植物的球茎，球茎状物，电灯泡
forefather	n.	祖先，祖宗
funding	n.	资金
hold	n.	握住，掌握，控制，容纳
chew	v.	咀嚼，认真考虑
dominant	adj.	有统治权的，占优势的，支配的
extremely	adv.	极端地，非常地
involve	v.	包括，涉及
knob	n.	(门，抽屉等的)球形捏手，节，瘤，旋钮
notoriously	adv.	臭名远扬地，声名狼藉地，众人皆知地
patch	n.	片，临时电子线路
pioneer	n.	先驱，倡导者，先遣兵，先锋
	v.	开辟，开拓，开创
prior	adj.	优先的，在前的
statement	n.	声明，陈述，语句
thorough	adj.	彻底的，完全的，细心的

Phrases

air conditioning	空气调节装置
heavy duty	重型，重载
University of Pennsylvania	宾夕法尼亚州大学
hydrogen bomb	氢弹
team up	(使)结成一队，合作，协作

Abbreviations

ENIAC (Electronic Numerical Integrator and Calculator) 电子数字积分计算机

1.1.2　正文参考译文

四类计算机

　　计算机是根据指令接收输入、处理输入数据并产生信息的电子设备。计算机有四种类型：微型机、小型机、大型机和巨型机。

　　微型计算机，也被称为个人计算机，是可以放在桌面上的小的计算机。便携式微型机可以放入手提箱，甚至手掌中。微型机被用于家庭、学校及工业中。如今几乎每一领域都在使用微型机。

　　便携式计算机是正在迅速普及的一种微型机，它易于四处携带。便携式计算机有四种类型。

　　膝上电脑：其重量在 10～16 磅之间，供电方式可以是交流供电、电池供电或两者均可。

交流供电的膝上电脑重量在 12~16 磅之间。电池供电的膝上电脑的重量(包括电池)在 10~15 磅之间，可以用肩带背起来携带。图 1.1 是一个膝上电脑的例子。

笔记本个人电脑：其重量在 5~10 磅之间，可放入大多数公文包中，尤其适用于那些连接电源不方便的场所。笔记本电脑是如今最流行的便携式电脑。

超小型笔记本电脑：适合那些经常飞来飞去和将时间花在道路上的人使用。这些用户放弃了大尺寸的显示器和键盘，换来了重量的减轻。这种电脑重量在 2~6 磅之间，可以很容易地放入公文包中。

个人数字助手：它比超轻薄笔记本电脑还要小得多，其重量在 1~4 磅之间。典型的个人数字助手将钢笔输入、书写识别、个人编排工具和通信功能结合起来放入小包中。图 1.2 是一个 PDA 的例子。

小型计算机，也被称为中型机，是像书桌大小的机器。它们的处理速度和数据存储能力介于微型机和大型机之间。中型公司或大型公司的部门一般把它们用于特殊用途。例如，可以使用它们做研究或监视某一个生产过程。小型公司一般使用小型机进行日常的数据处理，比如说统计。

大型机是较大的计算机，放置在具有专线、空调的房间中，具有很快的处理速度和很大的数据存储量。它们通常被一些大的组织机构(商业部门、银行、大学、政府机构)用来处理数以百万计的数据。例如，保险公司使用大型机处理数以百万计的保险客户的信息。

巨型机是非常大的机构进行研究工作的大容量专用计算机。这些应用包括石油勘探和世界范围的天气预报。一个巨型机的例子如图 1.3 所示。

一般来说，计算机的类型由下列 7 个因素决定：

CPU 的类型。微型计算机使用微处理器。较大的计算机趋向于使用由单独的高速复杂的零部件构成的 CPU。

CPU 能够使用的内存的总量。配备有大容量内存的计算机支持更复杂的程序，并且能同时容纳几个不同的程序运行。

存储设备的容量。较大的计算机系统趋向于配置较大容量的存储设备。

输出设备的速度。微机输出设备的速度用每秒钟能打印的字符数(cps)予以度量，通常为每秒几十个、几百个字符。较大计算机的输出设备的速度也较快，通常每分钟可打印几百或几千行。

用 mips(每秒钟百万条指令)度量的处理速度。在这里用术语"指令"来描述软件要求计算机完成的基本任务，并且标识受到影响的数据。较小计算机的处理速度为 7~40 mips。大型计算机的处理速度能达到 30~150 mips 或更多。巨型计算机的处理速度达 200 mips 以上。换句话说，大型计算机处理数据的能力要比微机快得多。

可以同时访问计算机的用户数量。大多数小型计算机某个时刻只能供单个用户使用，有些计算机可以同时由两个或三个用户访问，大型计算机则可同时由几百个用户访问。

计算机系统的价格。商用计算机系统的价格可低到 500 美元(一台微机)或者高达 1000 万美元(一台大型机)，巨型计算机则花费更多。

1.1.3 阅读材料参考译文

ENIAC

现代电子计算机的祖先通常被认为是 ENIAC，ENIAC 是电子数字积分计算机的简称，英文全称为 Electronic Numerical Integrator And Calculator。ENIAC 于 1943—1945 在宾夕法尼亚大学由两个教授——John Mauchly 和 24 岁的 J. Presper Eckert 制造。他们承诺可以建造一台机器代替所有的计算员(指雇佣来为军队的炮弹计算火力表的妇女)，因此得到了美国陆军部的资助。

ENIAC 占满了 20×40 英尺的房间，重 30 吨，并使用 1.8 万多个真空管。ENIAC 使用来自 IBM 的纸卡阅读器。工作时，ENIAC 是无声的，但我们知道它在工作，因为 18 000 个真空管每个都像灯泡一样产生废热，这些热量(174 000 瓦特热)意味着计算机只能运行在一个特殊设计的有自己的重型空调系统的房间。在图 1.4 的第一张照片中只能看到左边一半 ENIAC，右边的一半基本上是所看到内容的镜像。

为了修改 ENIAC 的程序，人们需要重新排列在图 1.4 中所看到的左边的连线，并设置右边的 3 000 个开关。对一个现代计算机编程，可以输入如下语句的程序：

$$\text{Circumference} = 3.14 * \text{diameter}$$

为了在 ENIAC 上执行此计算，人们需要重新安排大量的跳线，然后在巨大的旋钮墙上找到三个特殊旋钮，并将其设为 3、1 和 4。

其中最明显的问题是，该设计将需要 18 000 个真空管同时进行工作。真空管是极不可靠的。18 000 个真空管同时工作的想法被认为是不可能的，当时主要的真空管供应商——RCA，拒绝参加该项目(但是为了"战时合作"的利益仍然提供了真空管)。Eckert 认真设计电路，解决了真空管的可靠性问题。他非常专注，在选择应用于 ENIAC 的电线电缆类型时，先进行了一个实验。他把实验室的老鼠饿了几天，然后给它们所有可用电缆类型的样品，找出老鼠最不愿意吃的那一种。图 1.4 中是 ENIAC 中的一小部分真空管。

即使使用 18 000 个真空管，ENIAC 在某个时间也只能处理 20 个数据。ENIAC 基本时钟频率为 10 万赫兹。今天的家用电脑使用的时钟频率是十亿赫兹。从美国军队接受了 50 万美元的经费，ENIAC 的首要任务是计算是否有可能制造一个氢弹(原子弹是在战争期间完成的，因此比 ENIAC 早)。最初的问题在 ENIAC 运行仅需 20 秒，并与机械计算器工作四十小时后得出的答案进行核对。花费 6 个星期为 50 万纸卡打孔后，它宣布制造氢弹可行，ENIAC 并没有为人类带来好处。这个最初的 ENIAC 程序至今仍然是保密的。

后来 Eckert、Mauchly 和数学家约翰·冯·诺依曼联合一起设计 EDVAC，开创了存储程序计算机。

1.2 Computer Hardware

Computer hardware has four parts: the central processing unit (CPU) and memory, storage hardware, input hardware, and output hardware.

CPU The part of the computer that runs the program is known as the processor or central processing unit (CPU). In a microcomputer, the CPU is on a single electronic component, the microprocessor chip, within the system unit or system cabinet. The CPU itself has two parts: the control unit and the arithmetic-logic unit. In a microcomputer, these are both on the microcomputer chip.

The Control Unit The control unit tells the rest of the computer system how to carry out a program's instructions. It directs the movement of electronic signals between memory and the arithmetic-logic unit. It also directs these control signals among the CPU, input and output devices.

The Arithmetic-Logic Unit The arithmetic-logic unit, usually called the ALU, performs two types of operations—arithmetic and logical. Arithmetic operations are, as you might expect, the fundamental math operations: addition, subtraction, multiplication, and division. Logical operations consist of comparisons. That is, two pieces of data are compared to see whether one is equal to, less than, or greater than the other.

Memory Memory is also known as primary storage, internal storage, and it temporarily holds data, program instructions, and information. One of the most important facts to know about memory is that part of its content is held only temporarily. In other words, it is stored only as long as the computer is turned on. When you turn the machine off, the content will immediately vanish. The stored content in memory is volatile and can vanish very quickly.

Storage Hardware [1]The purpose of storage hardware is to provide a means of storing computer instructions and data in a form that is relatively permanent, that is, the data will not be lost when the power is turned off—and easy to retrieve when needed for processing. There are four kinds of storage hardware: floppy disks, hard disks, optical disk, and magnetic tape.

Floppy Disks Floppy disks are also called diskettes, flexible disks, floppies, or simply disks. The plastic disk inside the diskette cover is flexible, not rigid. They are flat, circular pieces of mylar plastic that rotate within a jacket. Data and programs are stored as electromagnetic charges on a metal oxide film coating the mylar plastic.

Hard Disks Hard disks consist of metallic rather than plastic platters. They are tightly sealed to prevent any foreign matter from getting inside. Hard disks are extremely sensitive instruments. The read-write head rides on a cushion of air about 0.000 001 inch thick. It is so thin that a smoke particle, fingerprint, dust, or human hair could cause what is known as a head crash. A head crash happens when the surface of the read-write head or particles on its surface contact the magnetic disk surface. A head crash is a disaster for a hard disk. It means that some or all of the data on the disk is destroyed. Hard disks are assembled under sterile conditions and sealed from impurities within their permanent containers.

Optical Disks Optical disks are used for storing great quantities of data. An optical disk can hold 650 megabytes of data—the equivalent of hundreds of floppy disks. Moreover, an optical disk makes an immense amount of information available on a microcomputer. In optical-disk technology, a laser beam alters the surface of a plastic or metallic disk to represent data. To read the data, a laser scans these areas and sends the data to a computer chip for conversion.

Magnetic Tape Magnetic tape is an effective way of making a backup, or duplicate, copy of your programs and data. We mentioned the alarming consequences that can happen if a hard disk suffers a head crash. You will lose some or all of your data or programs. Of course, you can always make copies of your hard-disk files on floppy disks. However, this can be time-consuming and may require many floppy disks. Magnetic tape is sequential access storage and can solve the problem mentioned above.

Input Hardware   Input devices take data and programs, and people can read or understand and convert them to a form the computer can process. This is the machine-readable electronic signals of 0s and 1s. Input hardware is of two kinds: keyboard entry and direct entry.

Keyboard Entry   Data is input to the computer through a keyboard that looks like a typewriter keyboard but has additional keys. In this way, the user typically reads from an original document called the source document. The user enters that document by typing on the keyboard.

Direct Entry   Data is made into machine-readable form as it is entered into the computer, no keyboard is used. Direct entry devices may be categorized into three areas: pointing devices (for example, mouse, touch screen, light pen, digitizer), scanning devices (for example, image scanner, fax machine, bar-code reader), and voice-input devices.

Output Hardware   Output devices convert machine-readable information into people-readable form. Common output devices are monitors, printers, plotters, and voice output.

Monitors   Monitors are also called display screen or video display terminals. Most monitors that sit on desks are built in the same way as television sets, and these monitors are called cathode-ray tubes. Another type of monitor is flat-panel display, including liquid-crystal display (LCD), electroluminescent (EL) display and gas-plasma display. An LCD does not emit light of its own. Rather, it consists of crystal molecules. [2]An electric field causes the molecules to line up in a way that alters their optical properties. Unfortunately, many LCDs are difficult to read in sunlight or other strong light. A gas-plasma display is the best type of flat screen. Like a neon light bulb, the plasma display uses a gas that emits light in the presence of an electric current.

Printers   There are four popular kinds of printers: dot-matrix, laser, ink-jet, and thermal.

Dot-Matrix Printer   Dot-matrix printers can produce a page of text in less than 10 seconds and are highly reliable. They form characters or images using a series of small pins on a print head. The pins strike an inked ribbon and create an image on paper. Printers are available with print heads of 9, 18, or 24 pins. One disadvantage of this type of printer is noise.

Laser Printer   The laser printer creates dot-like images on a drum, using a laser beam light source. [3]The characters are treated with a magnetically charged ink-like toner and then are transferred from drum to paper. A heat process is used to make the characters adhere. The laser printer produces images with excellent letter and graphics quality.

Ink-Jet Printer   An ink-jet printer sprays small droplets of ink at high speed onto the surface of the paper. This process not only produces a letter-quality image but also permits printing to be done in a variety of colors.

Thermal Printer   A thermal printer uses heat elements to produce images on heat-sensitive paper. Color thermal printers are not as popular because of their cost and the requirement of specifically treated paper. They are a more special use printer that produces near photographic output. They are widely used in professional art and design work where very high quality color is essential.

Plotters Plotters are special-purpose output devices for producing bar charts, maps, architectural drawings, and even three-dimensional illustrations. Plotters can produce high-quality multicolor documents and also documents that are larger in size than most printers can handle. There are four types of plotters: pen, ink-jet, electrostatic, and direct imaging.

Voice-Output Devices Voice-output devices make sounds that resemble human speech but actually are pre-recorded vocalized sounds. Voice output is used as a reinforcement tool for learning, such as to help students study a foreign language. It is used in many supermarkets at the checkout counter to confirm purchases. Of course, one of the most powerful capabilities is to assist the physically challenged.

Words

adhere	v.	黏附，附着，坚持
architectural	adj.	建筑上的，建筑学的
arithmetic	n.	算术，运算
assemble	v.	集合，聚集，装配
cabinet	n.	橱柜，机箱
checkout	n.	检验，收款处
chip	n.	芯片
crystal	adj.	结晶状的
	n.	晶体
cushion	n.	垫子，软垫，衬垫
droplet	n.	小滴
duplicate	n.	复制品，副本
electromagnetic	adj.	电磁的
flexible	adj.	柔韧的，易弯曲的
impurity	n.	杂质，混杂物，不洁，不纯
immense	adj.	极广大的，无边的
megabyte	n.	兆字节
metallic	adj.	金属的
molecule	n.	分子
multiplication	n.	乘法，增加
mylar	n.	聚酯薄膜
neon	n.	氖，氖光灯，霓虹灯
optical	adj.	光学的，眼的，视力的
oxide	n.	氧化物
plasma	n.	等离子体，等离子区
plotter	n.	绘图仪
reinforcement	n.	增援，加强，加固，援军

sensitive	*adj.*	敏感的，灵敏的，感光的
sterile	*adj.*	贫脊的，不育的，消过毒的，无菌的
spray	*v.*	喷射，喷溅
temporarily	*adv.*	暂时地，临时地
thermal	*adj.*	热的，热量的
toner	*n.*	调色剂，调色者，碳粉
vocalize	*v.*	成为有声
volatile	*adj.*	挥发性的，可变的，不稳定的

Phrases

dot-matrix printer	点阵式打印机
head crash	磁头划伤
ink-jet printer	喷墨式打印机
laser printer	激光打印机
line up	排列，(使)排成行，(使)对齐
thermal printer	热敏式打印机

Abbreviations

ALU(Arithmetic-Logic Unit)	算术–逻辑单元
EL(Electroluminescent)	电致发光
LCD(Liquid-Crystal Display)	液晶显示器

Notes

[1] 例句：The purpose of storage hardware is to provide a means of storing computer instructions and data in a form that is relatively permanent, that is, the data is not lost when the power is turned off—and easy to retrieve when needed for processing.

分析：句中 means 是"方法"的意思。when needed for processing 是状语从句 when they are needed for processing 的省略形式。为了简洁起见，有的状语从句(如时间状语从句、条件状语从句等)有时可省略从句的主语和部分谓语(尤其是当从句主语与主句主语一致，且从句谓语包括有动词 be 时)。例如，He kept silent when (he was) asked why he was late. 当被问及为什么迟到时，他一言不发。

译文：存储硬件的作用是以一种相对持久的方式提供存储计算机指令和数据的方法，即当切断电源时不会丢失数据，且当需要处理数据时又容易恢复。

[2] 例句：An electric field causes the molecules to line up in a way that alters their optical properties.

分析：句中 line up 意思是"排成行"，that alters their optical properties 是定语从

句，修饰 way。

译文：电场使得这些分子排成一行，这种排行改变着它们的光学特性。

[3] 例句：The characters are treated with a magnetically charged ink-like toner and then are transferred from drum to paper.

分析：句中 treat with 意思是"用……处理"，then 后面省略了主语 the characters。

译文：字符先被用磁化的带电的像墨一样的碳粉处理，然后被从磁鼓传送到纸上。

Exercises

Ⅰ. Put "true" or "false" in the brackets for the following statements according to the passage.

1. (　) The CPU is the processor of a computer.
2. (　) The memory of a computer can hold data and information permanently.
3. (　) Diskettes are flexible, flat, circular pieces of mylar plastic that rotate within a jacket, and they can be bent easily.
4. (　) Hard disks are instruments extremely sensitive to pollution, so magnetic tape is an effective compensation for making a backup of your programs and data.
5. (　) Input devices accept people readable data and programs, and convert them to machine readable form; while output devices reverse the process.
6. (　) Direct entry doesn't need keyboard to input data and information into computer.
7. (　) Mouse, touch screen, light pen, digitizer and plotter are all direct entry devices.
8. (　) Thermal printer must use specialized paper.
9. (　) Laser printer can produce images with excellent letter and graphics quality, and it can also print large size documents.
10. (　) Voice output devices can mimic human speech immediately.

Ⅱ. Fill in the blanks according to the passage.

1. The CPU itself has two parts: _____ and _____.
2. The ALU performs two operations: _____ and _____.
3. The basic math operations are: _____, _____, _____, _____.
4. Image scanner, fax-machine and bar-code reader are all _____.
5. LCD, EL and gas-plasma are all _____ displays.
6. Output devices convert _____ information into _____ form.
7. LCDs are difficult to read in _____ or other _____.
8. One disadvantage of dot-matrix printer is _____.
9. Plotters are special-purpose output devices, for producing _____, _____, _____, and even _____ illustrations.
10. Voice-output devices make sounds that resemble human speech but actually are _____.

Ⅲ. Translate the following words and expressions into Chinese.

1. arithmetic-logic unit
2. volatile
3. optical disk
4. electromagnetic charges
5. time-consumig
6. direct entry

7. voice-input
8. inked ribbon
9. letter-quality
10. heat sensitive

1.2.1 Reading Material

Flash Memory

Flash memory is a non-volatile computer memory that can be electrically erased and reprogrammed. It is a technology that is primarily used in memory cards and USB flash drives (as shown in Figure 1.5) for general storage and transfer of data between computers and other digital products. It is a specific type of EEPROM (Electrically Erasable Programmable Read-Only Memory) that is erased and programmed in large blocks; in early flash the entire chip had to be erased at once. Flash memory costs far less than byte-programmable EEPROM and therefore has become the dominant technology wherever a significant amount of non-volatile, solid state storage is needed. Example applications include PDAs (personal digital assistants), laptop computers, digital audio players, digital cameras and mobile phones. It has also gained popularity in the game console market, where it is often used instead of EEPROMs or battery-powered SRAM for game save data.

Figure 1.5 A USB flash drive. The chip on the left is the flash memory. The microcontroller is on the right

Flash memory is non-volatile, which means that no power is needed to maintain the information stored in the chip. In addition, flash memory offers fast read access times (although not as fast as volatile DRAM memory used for main memory in PCs) and better kinetic shock resistance than hard disks. These characteristics explain the popularity of flash memory in portable devices. Another feature of flash memory is that when packaged in a "memory card", it is enormously durable, being able to withstand intense pressure, extremes of temperature, and even immersion in water.

Although technically a type of EEPROM, the term "EEPROM" is generally used to refer specifically to non-flash EEPROM which is erasable in small blocks, typically bytes. Because erase cycles are slow, the large block sizes used in flash memory erasing give it a significant speed advantage over old-style EEPROM when writing large amounts of data.

Flash memory was invented by Dr. Fujio Masuoka while working for Toshiba circa 1980. According to Toshiba, the name "flash" was suggested by Dr. Masuoka's colleague, Mr. Shoji Ariizumi, because the erasure process of the memory contents reminded him of a flash of a camera. Dr. Masuoka presented the invention at the IEEE 1984 International Electron Devices Meeting (IEDM) held in San Francisco, California.

Words

circa	*prep.*	大约
console	*n.*	[计]控制台
durable	*adj.*	持久的，耐用的
enormously	*adv.*	非常地，巨大地
immersion	*n.*	沉浸
kinetic	*adj.*	(运)动的，动力(学)的
maintain	*v.*	维持，维修，继续
Toshiba	*n.*	日本东芝公司

Phrases

EEPROM (Electrically Erasable Programmable Read-Only Memory)　　电可擦除只读存储器

1.2.2　正文参考译文

计算机硬件

计算机硬件有四个组成部分：中央处理器和内存、存储硬件、输入硬件和输出硬件。

CPU　计算机运行程序的部分被称为处理器或中央处理单元。在微型计算机中，CPU在系统单元或系统机箱内的单独电子元件，即微处理器芯片上。CPU本身具有两个部分：控制单元和算术—逻辑单元。在微型计算机中，这两个部分都在微型机芯片上。

控制单元　控制单元告诉计算机系统的其他部分如何完成程序指令。它指挥着电子信号在内存和算术—逻辑单元之间的移动。它也控制着CPU和输入/输出设备之间的控制信号。

算术—逻辑单元　通常被称为ALU，完成两类运算——算术和逻辑。算术运算是基本的数学运算：加、减、乘、除。逻辑运算是由比较(运算)构成的。也就是说，对两个数据进行比较，以看其中一个是否是等于、小于或大于另外一个。

内存　内存也被称为主存储器、内部存储器，它临时存储数据、程序指令和信息。关于内存需要重点了解的是它所保存的内容只是临时的。换句话说，这些内容只有在计算机开着时才能被保存。当关闭机器时，其内容会立即消失。在内存中所存储的信息是不稳定的并会很快消失。

存储硬件　存储硬件的作用是以相对持久的方式提供存储计算机指令和数据的方法，即当切断电源时不会丢失数据，且当需要处理数据时又容易恢复。目前有四种存储硬件：软盘、硬盘、光盘和磁带。

软盘　软盘又被称为软磁盘、可弯曲磁盘、软盘或简单地称为磁盘。在磁盘封套内是柔韧的圆形聚酯塑料盘片，它们在封套内旋转。程序和数据以电磁荷的形式存储在聚酯塑料片表面的金属氧化物薄膜上。

硬盘　硬盘由金属盘片而不是塑料盘片组成。它们被紧紧地密封起来，以防止外界东西进入。硬盘是非常灵敏的设备。读写头浮在大约0.000 001英寸厚的空气气垫上。空气垫非常薄，以至于烟粒、指印、灰尘或者头发都可能引起磁头划伤。磁头划伤对于硬盘来讲是灾难性的，它意味着磁盘上的数据部分或全部丢失。硬盘在无菌条件下安装并且密封在远离杂质的永久的容器内。

光盘 光盘用于存储大量的数据。一个光盘可能容纳 650 兆字节的数据——相当于数以百计的软盘。并且，光盘使得大量的信息可用于微机上。在光盘技术中，激光束改变塑料或金属盘的表面来表示数据。为了读取数据，激光扫描这些区域并且将这些数据送给计算机芯片以便转换。

磁带 磁带可有效地备份(即复制、拷贝)程序和数据。我们曾提到如果硬盘磁头划伤，就会产生令人担忧的结果，因为这将会丢失部分或全部的程序或数据。当然，也可以将硬盘上的文件拷贝到软盘上。但这样很费时，并且需要很多张软盘。磁带是顺序访问存储的，能够解决上面所提到的问题。

输入硬件 输入硬件接收人们能读懂的程序和数据，并将其转换为计算机能处理的形式，这就是机器可读的电子信号 0 和 1。输入硬件有键盘输入和直接输入两种。

键盘输入 数据通过形似打字机键盘但有附加键的键盘输入到计算机。用这种方式，用户一般读取被称为是源文件的初始文件，通过在键盘上打字输入文件。

直接输入 当数据输入到计算机时，是以机器可读懂的形式输入的，不需要键盘。直接输入设备分成三类：指针设备(如鼠标、触摸屏、光笔、数字化仪)、扫描设备(如图像扫描仪、传真机、条形码读器)和声音输入设备。

输出硬件 输出设备将机器可读的信息转换为人类可读的形式。一般的输出设备有监视器、打印机、绘图仪和声音输出设备。

监视器 监视器也被称为屏幕显示或视频显示终端。大多数放在桌面上的监视器的制作方法同电视机一样，它们被称为是阴极射线管。另一类监视器是平板显示器，包括液晶显示器、光电发光显示器和等离子显示器。液晶显示器自己不发射光，相反，是由晶体分子组成，电场使得这些分子排成一行，这种排行改变着它们的光学特性。遗憾的是，许多液晶显示器在太阳光或其他强光下很难读到。等离子显示器是平板显示器中最好的一种。与氖光灯泡一样，等离子显示器在电流存在的情况下使用一种发光的气体。

打印机 目前普遍使用的打印机有四种：点阵式、激光式、喷墨式和热敏式。

点阵式打印机 它能在不到几秒的时间内打印一页文本并且非常稳定。点阵式打印机利用在打印头上的一系列小针来形成字符或图像，这些针击打喷墨的色带并在纸上产生图像。目前有 9 针、18 针和 24 针的打印机，这种打印机的缺点是它的噪声较大。

激光打印机 它使用激光束光源在磁鼓上产生小点一样的图像，并用磁化的带电的像墨一样的碳粉处理这些字符，然后从磁鼓传送到纸上，再使用热处理过程使这些字符粘贴。激光打印机打印的图像字符清晰，图像质量高。

喷墨式打印机 它能以很快的速度将小点状墨汁喷到纸面上。该类打印机不仅可用于印刷高质量的图像，而且能打印彩色图像。

热敏式打印机 它使用热元素在热感应纸上产生图像。由于价格高并需要特殊处理的纸张，彩色热敏打印机还不是很普及。热敏式打印机能产生逼真的输出。它们被广泛应用在要求高质量彩色输出的专业艺术设计工作中。

绘图仪 绘图仪是特殊用途的输出设备，用于产生条形图、地图、建筑绘图，甚至三维图表。绘图仪可以输出高质量的多种色彩的文档，并且文档的尺寸比大多数打印机能处理的大。目前有四种类型的绘图仪：钢笔、喷墨、静电和直接图像。

声音输出设备 声音输出设备可以发出类似于人类说话的声音，但实际上声音是事先被录制的。声音输出作为强化工具被用于辅助学习，例如帮助学生学习外语。它还被用于许多超市的收款台来确认购买。当然，它最强大的功能是帮助残障者。

1.2.3 阅读材料参考译文

闪存

闪存(Flash memory)，是一种以电子方式擦除和重写的非易失性的存储器。这种技术主要用于存储卡与 USB 闪存驱动器(图 1.5 所示)，进行一般性资料的储存，以及在计算机与其他数字产品间交换传输资料。闪存是一种类型特殊的、以大区块擦除与重写的EEPROM(电可擦除可编程的只读存储器)，早期的闪存只能对整个芯片进行擦除。闪存的成本远比以字节为单位擦写的 EEPROM 低，其技术主要用于需要大量非易失性固态存储的场合。PDA(个人数字助手)、笔记本电脑、数字随身听、数字相机与手机上均可见到闪存。此外，闪存还较好地获得了游戏控制台的市场，借以取代储存游戏资料用的 EEPROM 或带有电池的 SRAM。

闪存是非易失性的，也就是说没有电源也可以使得存储在芯片上的信息不被丢失。另外，闪存提供快速读取(但没有 PC 主存中使用的易失性 DRAM 那么快)，而且与硬盘相比，闪存具有更佳的动态抗振性。这些特点使得闪存在便携式设备中得到广泛应用。闪存的另一个特性是：它被制成存储卡后非常耐用，可以承受高压力与极端的温度，甚至浸在水中也不会损坏。

虽然闪存在技术上属于 EEPROM，但是"EEPROM"通常特指非快闪式、以小区块方式清除的 EEPROM，它们的清除单位是字节。老式的 EEPROM 擦除循环相当缓慢，相比之下以大区块方式擦除的闪存在写入大量资料时则具有明显的速度优势。

闪存是 Fujio Masuoka 博士于 1980 年在为东芝公司工作时发明的。根据东芝公司的说法，Fujio Masuoka 博士的同事 Shoji Ariizumi 先生将其命名为"闪"(flash)，是因为这种新存储器删除内容的过程使他想起照相机的闪光灯。Masuoka 博士于 1984 年在美国加州旧金山举行的 IEEE 国际电子设备会议上介绍了该项发明。

1.3 System Software

Software refers to computer programs. Programs are the instructions that tell the computer how to process data into the form you want. There are two kinds of software: system software and application software.

System software is a collection of programs that enables application software to run on a computer system's hardware devices, it is background software and includes programs that help the computer manage its own internal resources.

Application software is a specialized program that enables the user to accomplish specific tasks.

In this text, we mainly discuss system software.

System software consists of four kinds of programs: bootstrap loader, diagnostic routines, basic input-output system and operating system. Among these four parts, the operating system is we most concerned with, which helps manage computer resources. Most important operating systems are Windows, Windows NT, OS/2, Macintosh and Unix.

Windows Windows gets its name because of its ability to run multiple applications at the same time, each in its own window. Windows offers graphical user interface (GUI), presents the user with graphic images of computer functions and data. It provides a standard mechanism for copying or moving information from one program to another. This mechanism, called the Clipboard, means that information created in one context is instantly reusable in another, you don't need to reenter information or work with clumsy data-transfer utilities. Windows also has DDE (dynamic data exchange) and OLE (object linking and embedding) functions. In DDE two or more applications can be linked. This way, data created in one application is automatically entered into the others. OLE, like DDE, links data between applications. Additionally, OLE allows the application receiving the data to directly access the application that created the data.

Windows NT Windows NT is an operating system designed to run on a wide range of powerful computers and microcomputers. It is a very sophisticated and powerful operation system. Developed by Microsoft, Windows NT is not considered a replacement for Windows. [1]Rather, it is an advanced alternative designed for very powerful microcomputers and networks. Windows NT has two major advantages when compared to Windows.

Multiprocessing It is similar to multitasking except that the applications are run independently at the same time. For instance, you could be printing a word processing document and using a database management program at the same time. [2]With multitasking, the speed at which the document is printed is affected by the demands of the database management program. With multiprocessing, the demands of the database management program do not affect the printing of the document.

Networking In many business environments, workers often use computer to communicate with one another and to share software using a network. This is made possible and controlled by special system software. Windows NT has network capabilities and security checks built into the operating system. This makes network installation and use relatively easy.

OS/2 OS/2 stands for Operating System/2. It was developed jointly by IBM and Microsoft Corporation. OS/2 has many similarities with Windows NT. It is designed for very powerful microcomputers and has several advanced features. Some of its advantages over Windows NT include:

Minimum system configuration Like Windows NT, OS/2 requires significant memory and hard disk space. However, OS/2 requires slightly less.

Windows application Like Windows NT, OS/2 does not have a large number of application programs written especially for it. OS/2 can also run Windows programs, but it runs these programs slightly faster than Windows NT.

Common user interface Microcomputer application programs written specifically for Windows NT, as well as for OS/2, have consistent graphics interfaces. Across applications, the user is provided with similar screen displays, menus and operations. Additionally, OS/2 offers a consistent interface with mainframes, minicomputers and microcomputers.

Macintosh Operation System   The Macintosh Software, which runs only on Macintosh computers, offers a high-quality graphical user interface and is very easy to use. Apple Macintosh System 7.5 designed for Apple computers using Motorola's PowerPC microprocessor, is a significant milestone for Apple. It is a very powerful operating system like Windows NT and OS/2. System 7.5 has network capabilities and can read Windows and OS/2 files. It has several advantages.

Ease of use   The graphical user interface has made the Macintosh popular with many newcomers to computing. This is because it is easy to learn.

Quality graphics   Macintosh has established a high standard for graphics processing. This is a principal reason why the Macintosh is popular for desktop publishing. Users are easily able to merge pictorial and text materials to produce nearly professional-looking newsletters, advertisements, and the like.

Consistent interfaces   Macintosh applications have a consistent graphics interface. Across all applications, the user is provided with similar screen displays, menus and operations.

Multitasking   Like Windows, Windows NT and OS/2, the Macintosh System enables you to do multitasking. That is, several programs can run at the same time.

Communications between programs   The Macintosh system allows application programs to share data and commands with other application programs.

Unix   Unix was originally developed by AT&T for minicomputers and is very good for multitasking. It is also good for networking between computers. Unix initially became popular in industry because for many years AT&T licensed the system to universities for a nominal fee. It is popular among engineers and technical people. With the arrival of very powerful microcomputers, Unix is becoming a larger player in the microcomputer world. Unix can be used with different types of computer systems, that is, it is a portable operating system. It is used with microcomputers, minicomputers, mainframes and supercomputers. The other operating systems are designed for microcomputers and are not nearly as portable. It also has the advantages of multitasking, multiprocessing, multiuser and networking.

Words

accomplish	v.	完成，达到，实现
application	n.	申请，应用，应用程序
bootstrap	n.	引导程序
diagnostic	n.	特征，症状，[计]诊断程序
	adj.	诊断的，用于诊断的
dynamic	adj.	动态的，动力(学)
embed	v.	嵌套，装入
graphical	adj.	绘画的

initially	adv.	最初
interface	n.	界面接口，连接
license	v.	许可，准许
mechanism	n.	机械装置，机构，机制
merge	v.	归并，合并
multiple	adj.	多(倍、路、重、道、次)的
newsletter	n.	通信稿(业务通信，简讯)
nominal	adj.	名义上的，有名无实的，名字的
pictorial	adj.	图示的，图像的
refer	v.	提交，谈及，归诸于
security	n.	安全(性)，保密(性)，安全措施
specialized	adj.	专业的，专门的

Phrases

| bootstrap loader | 引导装入程序 |
| diagnostic routines | 诊断例程 |

Abbreviations

DDE(Dynamic Data Exchange)	动态数据交换
OLE(Object Linking and Embedding)	对象链接和嵌入
OS(Operating System)	操作系统

Notes

[1] 例句：Rather, it is an advanced alternative designed for very powerful microcomputers and networks.

分析：句中 rather(常与 or 连用)，意思是"更确切地说"。alternative 在这里是名词，意为"另一个可供选择的(东西或办法等)"，designed for very powerful microcomputers and networks 是过去分词短语作定语，修饰 alternative。

译文：确切地说，它是为功能非常强大的微型计算机和网络设计的一个高级的可供选择的操作系统。

[2] 例句：With multitasking, the speed at which the document is printed is affected by the demands of the database management program.

分析：at which the document is printed 是定语从句，修饰 speed。

译文：在多任务情况下，文件打印的速度会受到数据库管理程序要求的影响。

Exercises

Ⅰ. Put "true" or "false" in the brackets for the following statements according to the passage.

1. (　) System software is background software, that is, without it, computers can't work.
2. (　) System software has four kinds of programs: bootstrap loader, diagnostic routines,

basic input-output system, and operating system.

3. (   ) One computer can only run one kind of system software.

4. (   ) Windows provides a standard mechanism called Clipboard, and it can copy or move data easily.

5. (   ) DDE allows the application receiving the data to directly access the application that created the data.

6. (   ) Windows NT is designed for replacement of Windows.

7. (   ) OS/2 requires the same memory and hard disk space as Windows does.

8. (   ) Macintosh operating system can run on all kinds of computers.

9. (   ) Unix is a portable operating system, that is, it can be used in different kinds of computer systems.

10. (   ) Macintosh computers are designed to use Intel's microprocessor.

II. Fill in the blanks according to the passage.

1. _____ are the instructions that tell the computer how to process data into the form you want.

2. _____ is background software and includes programs that help the computer manage its own _____ resources.

3. System software consists of four kinds of programs: _____, _____, _____, _____.

4. Windows also has DDE (_____) and OLE (_____) functions.

5. _____ is similar to multitasking except that the applications are run independently and at the same time.

6. OS/2 was developed jointly by _____ and _____ Corporation.

7. The Macintosh Software, which runs only on _____ computers, offers a high-quality _____ and is very easy to use.

8. Unix can be used with different types of computer systems, that is, it is a _____ operating system.

9. _____ is a standard mechanism for copying or moving information from one program to another.

10. Unix can be used with different types of computer systems. It also has the advantages of _____, _____, _____ and _____.

III. Translate the following words and expressions into Chinese.

1. background software
2. internal resources
3. standard mechanism
4. significant milestone
5. network capabilities
6. security checks
7. bootstrap loader
8. diagnostic routines
9. data-transfer utilities
10. advanced alternative

1.3.1 Reading Material

Windows Vista

Windows Vista is a line of operating systems developed by Microsoft for use on personal computers, including home and business desktops, laptops, Tablet PCs, and media center PCs. Prior to its announcement on July 22, 2005, Windows Vista was known by its codename "Longhorn". Development was completed on November 8, 2006, over the following three months it was released in stages to computer hardware and software manufacturers, business customers, and retail channels. On January 30, 2007, it was released worldwide, and was made available for purchase and download from Microsoft's website. The release of Windows Vista came more than five years after the introduction of its predecessor, Windows XP, the longest time span between successive releases of Microsoft Windows desktop operating systems.

Windows Vista contains many changes and new features, including an updated graphical user interface and visual style dubbed Windows Aero, improved searching features, new multimedia creation tools such as Windows DVD Maker, and redesigned networking, audio, print, and display sub-systems. Vista also aims to increase the level of communication between machines on a home network, using peer-to-peer technology to simplify sharing files and digital media between computers and devices. Windows Vista includes version 3.0 of the .NET Framework, which aims to make it easier for software developers to write applications than with the traditional Windows API.

Microsoft's primary stated objective with Windows Vista, however, has been to improve the state of security in the Windows operating system. One common criticism of Windows XP and its predecessors has been their commonly exploited security vulnerabilities and overall susceptibility to malware, viruses and buffer overflows. In light of this, Microsoft chairman Bill Gates announced in early 2002 a company-wide "Trustworthy Computing initiative" which aims to incorporate security work into every aspect of software development at the company. Microsoft stated that it prioritized improving the security of Windows XP and Windows Server 2003 above finishing Windows Vista, thus delaying its completion.

While these new features and security improvements have garnered positive reviews, Vista has also been the target of much criticism and negative press. Criticism of Windows Vista has targeted its high system requirements, its more restrictive licensing terms, the inclusion of a number of new digital rights management technologies aimed at restricting the copying of protected digital media, lack of compatibility with some pre-Vista hardware and software, and the number of authorization prompts for User Account Control.

As a result of these and other issues, Windows Vista had seen initial adoption and satisfaction rates lower than Windows XP. However, as of January 2009, it has been announced that Vista usage had surpassed Microsoft's pre-launch two-year-out expectations of achieving 200 million users by an estimated 150 million. As of the end of March 2009, Windows Vista is the second most widely used operating system in the world with a 23.42% market share; the most widely used is Windows XP with a 62.85% market share.

Words

dub	v.	[电影]配音
exploit	v.	开拓，开发，使用
framework	n.	构架，框架，结构
garner	v.	储藏，存放收集
incorporate	adj.	合并的，结社的，一体化的
	v.	合并，组成公司
malware	n.	恶意软件
predecessor	n	前辈，前任
prioritize	v.	对……区分优先次序
span	n.	跨度，跨距，范围
stage	n.	舞台，发展的进程，阶段或时期
susceptibility	n.	易感性
tablet	n.	写字板，书写板
vulnerability	n.	弱点，攻击

Abbreviations

API (Application Programming Interface)　　　[计]应用编程接口

1.3.2　正文参考译文

系统软件

　　软件指的是计算机程序。程序是告诉计算机如何将数据处理成人们想要的形式的指令。软件分为两种：系统软件和应用软件。
　　系统软件指能让应用软件在计算机系统硬件设备上运行的程序的集合，它是后台软件并且包括帮助计算机管理自己内部资源的程序。
　　应用软件是让用户能够完成特定任务的专门程序。
　　在这一节中，我们主要讨论系统软件。
　　系统软件由四种程序组成：引导装入程序、诊断例程、基本输入输出系统和操作系统。在这四种程序中，操作系统是人们最为关心的，它帮助管理计算机资源。最重要的操作系统有 Windows、Windows NT、OS/2、Macintosh 和 Unix。
　　Windows 之所以被称为窗口，是因为它具有能同时运行多个应用程序的能力，并且每一个程序都有自己的窗口。Windows 提供图形用户界面，为用户呈现计算机的功能和数据的图像。它提供用于将信息从一个程序拷贝或移动到另一个程序的标准机制。这个机制被称为剪贴板，也就是说在一个环境中产生的信息可以被另外一个环境立即使用，而用户不需要重新输入信息或使用繁琐的数据传送实用程序。Windows 还具有 DDE(动态数据交换)和 OLE(对象链接和嵌入)功能。在 DDE 中两个或更多的应用程序可以被链接。使用这种方法，在一个应用程序中产生的数据可以自动地进入其他的程序。OLE 类似于 DDE，用以在应用程序之间链接数据。另外，OLE 允许接收数据的应用程序直接访问建立这个数据的应用程序。

Windows NT　它是被设计运行于各种功能强大的计算机和微型计算机上的操作系统。它非常复杂且功能强大。Windows NT 由微软公司开发，并不是为了替代 Windows，确切地说，它是为功能非常强大的微型计算机和网络而设计的高级的可供选择的操作系统。和 Windows 相比，Windows NT 具有两个主要的优点：

多重处理　除了应用程序是同时独立运行之外，它类似于多任务。比如，我们可以在打印字处理文档的同时使用数据库管理程序。在多任务情况下，文件打印的速度会受到数据库管理程序的要求的影响。使用多重处理技术，就可以避免这个问题。

网络　在许多商业环境中，工作人员经常使用计算机相互进行交流，并且通过网络共享软件。这些都可以通过专门的系统软件实现和控制。Windows NT 具有网络功能和嵌入到操作系统内的安全检测功能，这使得通过网络安装和使用相对容易。

OS/2　OS/2 即操作系统/2。它是由 IBM 和微软公司联合开发的。OS/2 和 Windows NT 有许多相似之处。它也是设计给功能强大的微型计算机的，具有几个高级特点。相对于 Windows NT，OS/2 的优点包括：

最小的系统配置　像 Windows NT 一样，OS/2 需要大的内存和硬盘空间。然而，OS/2 需要的比 Windows NT 少。

窗口应用　OS/2 和 Windows NT 一样，专门为它编写的应用程序很少。OS/2 同样可以运行 Windows 程序，但它运行这些程序的速度比 Windows NT 快。

共同的用户界面　专门为 Windows NT 以及 OS/2 编写的微型计算机应用程序具有一致的图形界面。各种应用程序为用户提供类似的屏幕显示、菜单和操作。此外，OS/2 为大型机、小型机和微型机提供统一的界面。

Macintosh 操作系统　它是仅运行在 Macintosh 计算机上的 Macintosh 软件，提供高质量的图形用户界面并且易于使用。Apple Matintosh System 7.5 是为使用 Motorola 公司的 PowerPC 微处理器的 Apple 计算机设计的，对于 Apple 机来说是一个重要的里程碑。它像 Windows NT 和 OS/2 一样是功能强大的操作系统。System 7.5 具有网络功能并能读出 Windows 和 OS/2 文件。它具有以下几个优点。

容易使用　图形用户界面使得 Macintosh 在许多计算机新手中广受欢迎，这是因为它容易学习。

高质量的图像　Macintosh 已经建立了高标准的图像处理，这就是 Macintosh 在桌面印刷系统流行的主要原因。使用它，用户可以很容易地将图片和文字材料组织成比较专业的通信稿、广告等。

一致的界面　Macintosh 应用程序具有一致的图形界面。在所有的应用程序中，提供给用户相似的屏幕显示、菜单和操作。

多任务　像 Windows、Windows NT 和 OS/2 一样，Macintosh 系统能使用户实现多任务，即几个程序可同时运行。

程序之间的通信　Macintosh 系统允许应用程序之间共享数据和命令。

Unix　它起初是由 AT&T 为小型机开发的，非常适于执行多任务处理，它也适合于计算机之间的网络链接。由于多年来 AT&T 一直将系统以很少的费用提供给大学学生使用，因此 Unix 起初就在工业领域很普及。它在工程师和技术人员中也很普及。随着功能强大的微型计算机的到来，Unix 正成为微机世界的大玩家。Unix 可被用于不同类型的计算机系统，也就是说，它是可移植的操作系统。它可用于微型机、小型机、大型机和巨型机上，而其他的操作系统被设计用于微型机上且几乎不可移植。Unix 也具有多任务、多道处理、多用户和网络功能的优点。

1.3.3　阅读材料参考译文

Windows Vista

Windows Vista 是微软开发的系列操作系统之一，用于个人计算机，包括家用和商用桌面机、便携机、平板电脑、媒体中心电脑。在 2005 年 7 月 22 日发布前，Windows Vista 被称为"Longhorn."，2006 年 11 月 8 日完成开发，在接下来的三个月里对计算机硬件软件生产商、商业客户、零售渠道发布。2007 年 1 月 30 日，Windows Vista 在世界范围发布，大众可以购买，也可以从微软的网站下载。Windows Vista 距离上一版本 Windows XP 已有五年多的时间，这是历史上微软 Windows 桌面操作系统版本发行间隔时间最久的一次。

Windows Vista 有很多变化，增加了一些新功能，包含新版的图形用户界面和被称为"Windows Aero"的全新视觉风格、加强后的搜索功能、新的多媒体创作工具(如 Windows DVD Maker)，以及重新设计的网络、音频、输出和显示子系统。Vista 也使用点对点技术(peer-to-peer)提升电脑系统在家庭网络中的通讯能力，使得在不同电脑或装置之间共享文件与数字媒体内容变得更简单。Vista 包含.NET Framework 3.0，比起传统的 Windows API，Vista 致力于让软件开发者更容易写出应用程序。

微软 Windows Vista 的最初目标是改进 Windows 操作系统的安全状态。Windows XP 以及之前的版本受到批评的一大原因是系统经常出现安全漏洞，并且容易受到恶意软件、病毒或缓冲区溢出等问题的影响。据此，微软总裁比尔·盖茨在 2002 上半年宣布在全公司实行"可信赖的电脑行动"，这个活动的目的是让全公司各方面的软件开发部门一起合作，共同解决安全性的问题。微软宣称由于非常希望增进 Windows XP 和 Windows Server 2003 的安全性，因此延误了 Vista 的开发完成。

在这些新功能和安全性改善得到正面评价的同时，Windows Vista 也成为了许多批评和负面评价的目标。对 Windows Vista 的批评集中在它的高性能需求、更严格的授权条款、包含一些限制对受保护数字媒体进行复制的新数字版权管理技术、缺乏与之前硬件和软件的兼容、以及用户账户控制的授权提示次数等。

因为这样和那样的问题，Windows Vista 最初的采用率和满意率都低于 Windows XP。无论如何，据公布，2009 年 1 月 Vista 的用量已经超过发布前微软预计的 2 年达到 2 亿用户的目标，用量大约 3.5 亿。截至 2009 年 3 月底，Windows Vista 是世界第二个被广泛使用的操作系统，占市场份额的 23.42%。使用最多的是 Windows XP，占市场份额的 62.85%。

1.4　Application Software

Application software might be described as end-user software. Application software performs useful work on general-purpose tasks such as word processing and cost estimating. There are certain general-purpose programs that are widely used in nearly all career areas. They are word processing, electronic spreadsheets, graphic programs and so on. They are also called basic tools and have some common features.

Insert Point The insert point or cursor shows you where you can enter data next. Typically , it is a blinking vertical bar on the screen. You can move it around using a mouse or the directional arrow keys on many keyboards.

Menus Almost all software packages have menus. Typically, the menus are displayed in a menu bar at the top of the screen. When one of these is selected, a pull-down menu appears. This is a list of commands associated with the selected menu.

Help For most applications, one of the menus on the menu bar is Help. When selected, the Help options appear. [1]These options typically include a table of contents, a search feature to locate reference information about specific commands, and central options to move around.

Button Bars Button bars typically are below the menu bar. They contain icons or graphic representations for commonly used commands. This offers the user a graphic approach to selecting commands. [2]It is an example of a graphic user interface in which graphic objects rather than menus can be used to select commands.

Dialog Box Dialog boxes frequently appear after selecting a command from a pull-down menu. These boxes are used to specify additional command options.

Scroll Bars Scroll bars are usually located on the right and/or the bottom of the screen. They enable you to display additional information not currently visible on the screen.

WYSIWYG Pronounced "wizzy-wig", WYSIWYG stands for "What You See Is What You Get". This means that the image on the screen display looks the same as the final printed document. Application programs without WYSIWYG cannot always display an exact representation of the final printed document. The WYSIWYG feature allows the user to preview the document's appearance before it is printed out.

Function Keys Function keys are labeled F1, F2 and so on. These keys are positioned along the left side or along the top of the keyboard. They are used for commands or tasks that are performed frequently, such as underlining. These keys do different things in different software packages.

Now let's introduce respectively the most common used application software: word processing and spreadsheets.

Word processing software is used to create, edit, save, and print documents. Documents can be any kind of text material. With word processing, you view the words you type on a monitor instead of on a piece of paper. After you finish your typing, save your words on diskettes or hard disk, and print the results on paper.

The beauty of this method is that you can make changes or corrections—before printing out the document. Even after your document is printed out, you can easily go back and make changes. You can then print it out again. Want to change a report from double spaced to single spaced?

Alter the width of the margins on the left and right? Delete some paragraphs and add some others from yet another document? A word processor allows you to do all of them with ease. Indeed, deleting, inserting, and replacing—the principal correcting activities—can be done just by pressing keys on the keyboard. Popular word processing software are Word, WPS and so on. They have some common features.

Word Wrap and the Enter Key　One basic word processing feature is word wrap. When you finish a line, a word processor decides for you and automatically moves the insertion point to the next lines. To begin a new paragraph or leave a blank line, you press the Enter key.

Search and Replace　A search or find command allows you to locate any character, word, or phrase in your document. When you search, the insertion point moves to the first place the item appears. If you want, the program will continue to search for all other locations where the item appears. The replace command automatically replaces the word you search for with another word. The search and replace commands are useful for finding and fixing errors.

Cut, Copy, and Paste　With a word processor, you select the portion of text to be moved by highlighting it. Using either the menu or button bar, choose the command to cut the selected text. The selected text disappears from your screen. Then move the insertion point to the new location and choose the paste command to reinsert the text into the document. In a similar manner, you can copy selected portions of text from one location to another.

Spreadsheet　A spreadsheet is an electronic worksheet used to organize and manipulate numbers and display options for analysis. Spreadsheets are used by financial analysts, accountants, contractors, and others concerned with manipulating numeric data. Spreadsheets allow you to try out various "what-if" kinds of possibilities. That is a powerful feature. You can manipulate numbers by using stored formulas and calculate different outcomes.

A spreadsheet has several parts. The worksheet area of the spreadsheet has letters for column headings across the top. It also has numbers for row headings down the left side. The intersection of a column and row is called a cell. The cell holds a single unit of information. The position of a cell is called the cell address. For example, "A1" is the cell address of the first position on a spreadsheet, the topmost and leftmost position. A cell pointer—also known as the cell selector—indicates where data is to be entered or changed in the spreadsheet. The cell pointer can be moved around in much the same way that you move the insertion pointer in a word processing program. Excel is the most common spreadsheet software. It has some common features of spreadsheet programs.

Format　Label is often used to identify information in a worksheet, it is usually a word or symbol. A number in cell is called a value. Labels and values can be displayed or formatted in different ways. A label can be centered in the cell or positioned to the left or right. A value can be displayed to show decimal places, dollars or percent. The number of decimal positions can be altered, and the width of columns can be changed.

Formulas　One of the benefits of spreadsheets is that you can manipulate data through the use of formulas. Formulas are instructions for calculations. They make connections between numbers in particular cells.

Functions   Functions are built-in formulas that perform calculations automatically.

Recalculation   Recalculation or what-if analysis is one of the most important features of spreadsheets. If you change one or more numbers in your spreadsheet, all related formulas will recalculate automatically. Thus you can substitute one value for another in the cells affected by your formula and recalculate the results. For more complex problems, recalculation enables you to store long, complicated formulas and many changing values and quickly produce alternatives.

Words

accountant	n.	会计员(出纳员)
approach	n.	方法，手段
associate	v.	联合，联系，联想
blink	v.	闪烁
built-in	adj.	内置的，固定的
	n.	内置
contractor	n.	订约人，承包者，收敛部分，压力机
decimal	adj.	十进的
end-user	n.	最终用户
general-purpose	adj.	通用的
graphic	adj.	图形的
highlight	v.	使突出
intersection	n	相交
manipulate	v.	操作
margin	n.	页边空白
outcome	n.	结果，成果
package	n.	程序包，数据包
reference	n.	参考，引用
respectively	adv	各自地，分别地
spreadsheet	n.	电子表格，电子数据表，伸展表
underline	v.	在……下面画线
vertical	adj.	垂直(并排)的，纵向的
wrap	v.	换行

Phrases

pull-down menu	下拉菜单
software package	软件包

Abbreviations

WPS(Word Processing System)	字处理系统

WYSIWYG(What You See Is What You Get) 所见即所得

Notes

[1] 例句：These options typically include a table of contents, a search feature to locate reference information about specific commands, and central options to move around.

分析：句中 about specific commands 是介词短语作定语修饰 reference information，to move around 是动词不定式作定语修饰 options.

译文：帮助选项一般包含目录、查找有关特殊命令的说明信息的搜索功能和可以到处移动的集中选项。

[2] 例句：It is an example of a graphic user interface in which graphic objects rather than menus can be used to select commands.

分析：句中 rather than 的意思是"而不是"。

译文：工具栏就是图形用户界面的一个例子，在工具栏中通过图标而不是菜单来执行命令。

Exercises

Ⅰ. Put "true" or "false" in the brackets for the following statements according to the passage.

1. ( ) Application software includes some general-purpose programs that are widely used in nearly all career areas and some specialized programs that are used in special fields.

2. ( ) The insertion point or cursor indicates where you may enter data next.

3. ( ) Button bars and menus are all methods for users to pick up.

4. ( ) With WISYWIG, users can obtain representation of the exactly final results they want.

5. ( ) Function keys do the same things in different software programs.

6. ( ) Word wrap is a feature common to spreadsheet.

7. ( ) A cell in spreadsheet can hold several units of information.

8. ( ) Users can make connections among numbers in some cells through formulas.

9. ( ) Recalculation makes related formulas alter easily in a spreadsheet when one number in a cell is changed.

10. ( ) Spreadsheet programs are typically used to store and retrieve records quickly.

Ⅱ. Fill in the blanks according to the passage.

1. The insert point or _____ shows you where you can enter data next.

2. _____ menu is a list of commands associated with a selected menu.

3. Button bars typically are below the menu bar and contain _____ or _____ representations for commonly used commands.

4. The _____ feature allows the user to preview the document's appearance before it is printed out.

5. The search and replace commands are useful for _____ and _____ errors.

6. The intersection of a column and row in a spreadsheet is called a _____.

7. _____ are instructions for calculations.

8. The worksheet area of the spreadsheet has letters for column _____.

9. _____ enable you to display additional information not currently visible on the screen.

10. A search or find command allows you to locate any _____, _____, or _____ in your document.

III. Translate the following words and expressions into Chinese.

1. general-purpose
2. electronic spreadsheets
3. button bars
4. word wrap
5. pull-down menu
6. manipulate
7. recalculation
8. margin
9. text material
10. paste command

1.4.1 Reading Material

Software Engineer Job Description

A software engineer researches, designs and develops software systems to meet with clients' requirements. Once the system had been fully designed, software engineers then test, debug, and maintain the systems.

The software engineer job encompasses a fairly wide range of responsibilities. Smaller applications and system may employ just a few software engineers to manage the full lifecycle software development process. Generally, for most large scale applications, jobs are broken down into groups that focus on one area of the software lifecycle or just a specific area of the application or technology. For example, one system may employ a Software Architect, Design Engineer, Java Developer and Quality Assurance Engineer.

In today's market, jobs involving web services have become more common. Object oriented analysis and design have been common requirements for most business application design. Many of the responsibilities listed below are vague and general, focusing more on software engineering in a corporate setting. This does not encompass every possible software engineering responsibility and there are other specialized software engineering positions such as embedded software engineers.

Software engineers are sometimes referred to as computer programmers or software developers. Depending on the type of organization, software engineers can become specialists in either systems or applications.

Common Job Responsibilities for Software Engineer are as follows:

1. Full lifecycle application development;
2. Designing, coding and debugging applications in various software languages;
3. Software analysis, code analysis, requirements analysis, software review, identification of code metrics, system risk analysis, software reliability analysis;
4. Object-oriented Design and Analysis (OOA and OOD);

5. Software modeling and simulation;
6. Front end graphical user interface design;
7. Software testing and quality assurance;
8. Performance tuning, improvement, balancing, usability, automation;
9. Support, maintain and document software functionality;
10. Integrate software with existing systems;
11. Evaluate and identify new technologies for implementation;
12. Project Planning and Project Management;
13. Maintain standards compliance;
14. Implement localization or globalization of software.

Common IT Hardware, Software and Systems Knowledge are as follows.

C, C++, Java, .NET, Python, BEA WebLogic, WebSphere, J2EE, JBoss, ADO, Perl, HTML, JSP, JavaScript, Web services, SOAP, XML, ASP, JSP, PHP, MySQL, SQL Server, Oracle, UNIX, Linux, STL, XSLT, OWL, AJAX.

Words

architect	n.	建筑师
assurance	n.	确信,断言,保证,担保
compliance	n.	依从,顺从
encompass	v.	包围,环绕,包含或包括某事物
fairly	adv.	公正地,正当地,相当地,清楚地
focus	n.	中心,焦点,焦距
	v.	聚焦,注视,调焦,集中
identification	n.	辨认,鉴定
implement	n.	工具,器具
	v.	贯彻,实现
lifecycle	n.	生命周期,生活周期
metric	adj.	米制的,公制的
	n.	度量衡量标准
organization	n.	组织,机构,团体
responsibility	n.	责任,职责
simulation	n.	仿真,假装,模拟
tune	n.	曲调,和谐,合调
	v.	调音,调整
vague	adj.	含糊的,茫然的

Phrases

break down	分解
graphical user interface	图形用户界面

Software Engineer　　　　　　　　　　　　　软件工程师

1.4.2　正文参考译文

应用软件

应用软件可被描述为最终用户软件。应用软件完成有用的通用任务，例如字处理和价格评估。有些应用软件是被广泛用于几乎所有行业领域的通用程序，例如字处理、电子表格和图像程序等。它们也被称为基本工具，并且具有一些共同的特性。

插入点　插入点或光标显示标示着用户接下来可以输入数据的地方。一般来说，它是屏幕上闪动的竖直条。用户可以使用鼠标或键盘上的方向键移动它。

菜单　几乎所有的软件都有菜单。一般地，菜单显示在屏幕顶部的菜单栏中。选中其中一个菜单时，就会出现下拉式菜单，其中包括和所选菜单相关的一列命令。

帮助　对于大多数应用程序，菜单栏中都有一个"帮助"菜单。选中该菜单时，帮助选项出现。帮助选项一般包含目录、查找有关特殊命令的说明信息的搜索功能和可以到处移动的集中选项。

工具栏　工具栏一般位于菜单栏的下面。它们包含图标或普通命令的图像表示符。这就使得用户可以通过图像符号来选择命令。工具栏就是图形用户界面的一个例子，在工具栏中通过图标而不是菜单来执行命令。

对话框　通常在下拉式菜单中选择了一个命令之后就会出现对话框。这些对话框用来说明附加的命令选项。

滚动条　滚动条通常位于屏幕的右边或底部。它们通常能够用来帮助显示在当前屏幕上看不到的附加信息。

WYSIWYG　发"wizzy-wig"音，WYSIWYG 代表"所见即所得"，意即在屏幕上显示的图像和最终打印出来的文档是一样的。没有 WYSIWYG 的应用程序不能显示最终打印文档的精确表示。WYSIWYG 特性允许用户在打印文档之前预览文档的全貌。

功能键　功能键被标为 F1、F2 等，这些键位于键盘的左边或顶部。它们用于需经常完成的命令或任务，比如加下画线。这些键在不同的软件包内完成不同的任务。

接下来介绍最常使用的应用软件：字处理和电子表格。

字处理软件被用于建立、编辑、保存和打印文档。文档可以是任何类型的文本材料。使用字处理软件，可以在显示器浏览所输入的文字而不用打印在纸上。输入完毕后，可以将文字存放在软盘或硬盘上，并且可以将结果打印在纸上。

这种方式的精美之处在于用户在打印文档之前可以修改文档，即使文档已打印出来，也可以很容易地返回去进行修改，然后再打印出来。想把一个报告的行距从双倍改成单倍吗？想改变左右边距的宽度吗？想删除一些段落并插入其他文档的一些段落吗？字处理可以很容易地帮你完成这些事情。事实上，删除、插入和替换这几种基本的修改任务只要按键盘上的键就可以完成。目前流行的字处理软件有 Word、WPS 等，它们具有一些共同的特性。

字换行和 Enter 键　字处理的一个基本特点是换行。当完成一行时，字处理器就会决定并且自动将插入点移动到下一行。若要开始一个新段或空一段，可以按 Enter 键。

查找和替换　查找命令允许在文档中寻找任何字符、词汇或短语。查找时，插入点就会移动到所查项目首次出现的位置。如果需要，程序会继续寻找所查项目出现的所有地方。替换命令会自动用另外的字来替换所搜寻的那个字。寻找和替换命令对于发现和修改错误是非常有用的。

剪切、复制和粘贴 使用字处理，可以高亮显示并选择要移动的文本部分。通过菜单或工具栏，选择命令来剪切所选的文本，这样所选择的文本就会在屏幕上消失。然后移动插入点到新的位置，选择粘贴命令，即可把剪切的内容重新插入到文本中。用类似的方法，可以复制选中的文本到另外一个地方。

电子表格 电子表格是用于组织和管理数字并且显示选项以供分析的电子工作表，常由金融分析师、会计师、项目承包人以及其他和操纵数字数据有关的人员来使用。电子表格允许用户尝试各种可能的假设分析，这一特性很有用。可以通过使用存储的公式处理数字，并且计算出不同的结果。

电子表格具有几个部分，工作表区域的顶部有表示列标的字母，左边有表示行标的数字。行和列的交点被称为单元格。单元格存有单一的信息。单元格所在的位置被称为是单元格地址。例如，"A1"就是电子表格的首位置，即最顶部和最左边的位置。单元格指针(也被称为选择器)指示在表格中输入及修改数据的位置。单元格指针可以到处移动，其移动方式类似于在字处理程序中移动插入点的方式。Excel是最常见的电子表格软件，它具有一些电子表格程序共同的特性。

格式 标号往往用于标记工作表中的信息，通常是一个字或一个符号。单元格中的数字被称为是值。标号和值可以用不同的方式显示和格式化。标号可以在单元格内居中或居左、居右。值可被显示小数的位置、美元或百分数。我们可以改变小数位数，也可以改变列的宽度。

公式 电子表格的优点之一是可以通过使用公式来处理数据。公式是计算的指令，它们能使独立单元格内的数字之间建立联系。

函数 函数是自动完成计算的内部公式。

重新计算 重新计算或"what-if"分析是电子表格最重要的特性之一。如果改变了表格中的一个或多个数字，所有相关的公式将会自动地重新计算。这样就可以替换由公式改变的单元格内的值，并且重新计算结果。对于较复杂的问题，重新计算让用户能够存储长的、复杂的公式和许多改变的值，并且被很快地替换掉。

1.4.3 阅读材料参考译文

软件工程师职位描述

软件工程师研究、设计和开发软件系统，以满足客户需求。系统开发完成后，软件工程师要测试、调试和维护系统。

软件工程师工作涵盖了相当广泛的责任范围。规模较小的应用和系统可雇用几个软件工程师来管理全生命周期的软件开发过程。一般来说，在多数大规模的应用中，把工作分成几个小组，每个小组侧重于软件生命周期的某一阶段或只是应用或技术的某一领域。例如，开发一个系统可能需要雇用软件设计师、设计工程师、Java开发员和质量保证工程师。

在今天的市场上，与网络服务相关的工作变得更加常见。面向对象分析和设计已成为大多数企业设计应用程序的常见要求。下面列出的许多职责是不具体的，它们更注重企业设置中的软件工程。这并不包括一切可能的软件工程的责任，还有其他一些专门的软件工程的职位，如嵌入式软件工程师。

软件工程师，有时被称为计算机程序员或软件开发员。根据组织类型的不同，软件工程师可能成为系统软件或应用软件专家。

软件工程师的一般工作职责如下：

1. 全生命周期的应用程序开发；
2. 用各种软件语言设计、编写和调试应用程序；
3. 软件分析，代码分析，需求分析，软件审查，确定代码标准，系统风险分析，软件可靠性分析；
4. 面向对象的设计与分析(面向对象分析 OOA 和面向对象设计 OOD)；
5. 软件建模与仿真；
6. 前端图形用户界面设计；
7. 软件测试和质量保证；
8. 性能调整、改善、平衡、可用性、自动化；
9. 支持、维护软件并为软件编写文档的功能；
10. 将软件与现有系统集成；
11. 评估和确定新的实施技术；
12. 项目规划和项目管理；
13. 维持标准的遵守；
14. 实现软件的本地化或全球化。

通用的 IT 硬件、软件和系统知识如下。

C, C++, Java, .NET, Python, BEA WebLogic, WebSphere, J2EE, JBoss, ADO, Perl, HTML, JSP, JavaScript, Web services, SOAP, XML, ASP, JSP, PHP, MySQL, SQL Server, Oracle, UNIX, Linux, STL, XSLT, OWL, AJAX。

Chapter 2 Computer Network

2.1 Introduction to Computer Network

Computer network is a system connecting two or more computers. A computer network allows user to exchange data quickly, access and share resources including equipments, application software, and information.

Data communication systems are the electronic systems that transmit data over communication lines from one location to another. You might use data communications through your microcomputer to send information to a friend using another computer. You might work for an organization whose computer system is spread throughout a building, or even throughout the country or world. That is, all the parts—input and output units, processor, and storage devices—are in different places and linked by communications. Or you might use telecommunication lines—telephone lines—to tap into information located in an outside data bank. You could then transmit it to your microcomputer for your own reworking and analysis.

To attach to a network, a special-purpose hardware component is used to handle all the transmission. The hardware is called a network adapter card or network interface card (NIC), it is a printed circuit board plugged into a computer's bus, and a cable connects it to a network medium.

Communications networks differ in geographical size. There are three important types: LANs, MANs and WANs.

Local Area Networks Networks with computers and peripheral devices in close physical proximity—within the same building, for instance—are called local area networks (LANs). Linked by cable-telephone, coaxial, or fiber optic. LANs often use a bus form organization. In a LAN, people can share different equipments, which lower the cost of equipments. LAN may be linked to other LANs or to larger networks by using a network gateway. With the gateway, one LAN may be connected to the LAN of another office group. It may also be connected to others in the wide world, even if their configurations are different. Alternatively, a network bridge would be used to connect networks with the same configurations.

There is a newly development for LANs: WLAN. A wireless LAN (WLAN) is a flexible data communication system implemented as an extension to, or as an alternative for, a wired LAN within a building or campus. Using electromagnetic waves, WLANs transmit and receive data over the air, minimizing the need for wired connections. Thus, WLANs combine data connectivity with user mobility, and, through simplified configuration, enable movable LANs.

Over the recent several years, WLANs have gained strong popularity in a number of vertical markets, including the health-care, retail, manufacturing, warehousing, and academic arenas.

[1]These industries have profited from the productivity gains of using hand-held terminals and notebook computers to transmit real-time information to centralized hosts for processing. Today WLANs are becoming more widely recognized as a general-purpose connectivity alternative for a broad range of business customers.

Applications for Wireless LANs [2]Wireless LANs frequently augment rather than replace wired LAN networks—often providing the final few meters of connectivity between a backbone network and the mobile user. The following list describes some of the many applications made possible through the power and flexibility of wireless LANs:

- Doctors and nurses in hospitals are more productive because hand-held or notebook computers with wireless LAN capability deliver patient information instantly.
- Consulting or accounting audit engagement teams or small workgroups increase productivity with quick network setup.
- Network managers in dynamic environments minimize the overhead of moves, adds, and changes with wireless LANs, thereby reducing the cost of LAN ownership.
- Training sites at corporations and students at universities use wireless connectivity to facilitate access to information, information exchanges, and learning.
- Network managers installing networked computers in older buildings find that wireless LANs are a cost-effective network infrastructure solution.
- Retail store owners use wireless networks to simply frequent network reconfiguration.
- Trade show and branch office workers minimize setup requirements by installing preconfigured wireless LANs needing no local MIS support.
- Warehouse workers use wireless LANs to exchange information with central databases and increase their productivity.
- Network managers implement wireless LANs to provide backup for mission-critical applications running on wired networks.
- Senior executives in conference rooms make quicker decisions because they have real-time information at their fingertips.

The increasingly mobile user also becomes a clear candidate for a wireless LAN. Portable access to wireless networks can be achieved using laptop computers and wireless NICs. This enables the user to travel to various locations—meeting rooms, hallways, lobbies, cafeterias, classrooms, etc.—and still have access to their networked data. Without wireless access, the user would have to carry clumsy cabling and find a network tap to plug into.

Metropolitan Area Networks These networks are used as links between office buildings in a city. Cellular phone systems expand the flexibility of MAN by allowing links to car phones and portable phones.

Wide Area Networks Wide area networks are countrywide and worldwide networks. Among other kinds of channels, they use microwave relays and satellites to reach users over long distances. One of the most widely used WANs is Internet, which allows users to connect to other users and facilities worldwide.

Words

alternatively	adv.	二中择一地，换句话说
attach	v.	附上，连接
audit	v.	审计，会计检查，查账，核查
augment	v.	增大，增加
backbone	n.	构架，中心，中枢，主干线
cafeteria	n.	自助食堂
candidate	n.	选择物，候选人
clumsy	adj.	笨拙的
engagement	n.	约定
exchange	v.	交换，调换
facilitate	v.	易于，便于，助长
cost-effective	adj.	划算的
gateway	n.	网关
halfway	adj.	中途的，不彻底的
infrastructure	n.	下部结构，永久性基地，基础
lobby	n.	门廊，休息室
metropolitan	adj.	大城市的
mission	n.	使命，任务，代表团
	v.	派遣
mobility	n.	灵活性，移动性，可动性
overhead	adj.	过顶的，头上的，经常的
peripheral	n.	外部设备，辅助设备
plug	n.	插头，插塞
profit	v.	有利于，获益
proximity	n.	接近，近似，近程
retail	n.	零售
warehouse	n.	仓库

Abbreviations

LAN (Local Area Network)	局域网
MAN (Metropolitan Area Network)	城域网
MIS (Management Information System)	管理信息系统
NIC (Network Interface Card)	网络接口卡
WAN (Wide Area Network)	广域网
WLAN (Wireless Local Area Network)	无线局域网

Notes

[1] 例句：These industries have profited from the productivity gains of using hand-held terminals and notebook computers to transmit real-time information to centralized hosts for processing.

分析：句中 of using hand-held terminals and notebook computers to transmit real-time information to centralized hosts for processing 为介词短语修饰 gains，其中 to transmit real-time information to centralized hosts for processing 为 using hand-held terminals and notebook computers 的目的状语。此句虽然较长，但是一个简单句。

译文：这些行业通过手提终端和笔记本电脑将实时信息传送到中央主机进行处理，提高生产率，已获益匪浅。

[2] 例句：Wireless LANs frequently augment rather than replace wired LAN networks—often providing the final few meters of connectivity between a backbone network and the mobile user.

分析：often providing…是分词短语作 augment rather than replace 的伴随状语。

译文：无线局域网常常扩充而并非代替有线局域网的功能，它通常提供骨干网络和移动用户间最后几米的连接。

Exercises

Ⅰ. Put "true" or "false" in the brackets for the following statements according to the passage.

1. (　) A computer network is only connected by cable lines, such as telephone lines.
2. (　) The three types of network mainly differ in their geographical size.
3. (　) With a network gateway, a LAN can be connected to another LAN that has the same configuration.
4. (　) A network bridge and gateway are connectors that are used to connect two LANs or more.
5. (　) WLAN is a replacement for LAN.
6. (　) NIC is an electric circuit board that is a necessary hardware component of a computer network.
7. (　) WLANs transmit data through the air and they needn't NICs.
8. (　) A metropolitan area network connects two or more computers within a city.
9. (　) Internet is the most often used WAN of today.
10. (　) With WLANs, it is possible for users to connect with networks at any corner of the world.

Ⅱ. Fill in the blanks according to the passage.

1. A computer network allows user to _____ data quickly, _____ and _____ resources.
2. WLAN stands for _____.

3. Wireless LANs frequently _____ rather than _____ wired LAN networks.

4. _____ phone systems expand the flexibility of MAN by allowing links to car phones and portable phones.

5. Network managers in _____ environments _____ the overhead of moves, adds, and changes with wireless LANs, thereby reducing the _____ of LAN ownership.

6. Wide area networks are _____ and _____ networks.

7. Data communications systems are the electronic systems that transmit data over _____ from one location to another.

8. Senior executives in conference rooms make quicker decisions because they have _____ information at their fingertips.

9. A wireless LAN is a flexible data communication system implemented as an _____ to, or as an _____ for, a wired LAN within a building or campus.

10. A network bridge would be used to connect networks with the same _____.

III. Translate the following words and expressions into Chinese.

1. telecommunications lines
2. network interface card
3. geographical size
4. electromagnetic wave
5. gateway
6. hand-held
7. clumsy cabling
8. peripheral devices
9. information exchanges
10. flexibility

2.1.1 Reading Material

The 7 Layers of the OSI Model

The OSI, or Open System Interconnection model, defines a networking framework for implementing protocols in seven layers. Control is passed from one layer to the next, starting at the application layer in one station, proceeding to the bottom layer, over the channel to the next station and back up the hierarchy.

Application (Layer 7)	This layer supports application and end-user processes. Communication partners are identified, quality of service is identified, user authentication and privacy are considered, and any constraints on data syntax are identified. Everything at this layer is application-specific. This layer provides application services for file transfers, e-mail, and other network software services. Telnet and FTP are applications that exist entirely in the application level. Tiered application architectures are part of this layer.
Presentation (Layer 6)	The presentation layer works to transform data into the form that the application layer can accept. This layer formats and encrypts data to be sent across a network, providing freedom from compatibility problems. It is sometimes called the syntax layer.

Session (Layer 5)	This layer establishes, manages and terminates connections between applications. The session layer sets up, coordinates, and terminates conversations, exchanges, and dialogues between the applications at each end. It deals with session and connection coordination.
Transport (Layer 4)	This layer provides transparent transfer of data between end systems, or hosts, and is responsible for end-to-end error recovery and flow control. It ensures complete data transfer.
Network (Layer 3)	This layer provides switching and routing technologies, creating logical paths, known as virtual circuits, for transmitting data from node to node. Routing and forwarding are functions of this layer, as well as addressing, internetworking, error handling, congestion control and packet sequencing.
Data Link (Layer 2)	At this layer, data packets are encoded and decoded into bits. It furnishes transmission protocol knowledge and management and handles errors in the physical layer, flow control and frame synchronization. The data link layer is divided into two sub layers: the Media Access Control (MAC) layer and the Logical Link Control (LLC) layer. The MAC sub layer controls how a computer on the network gains access to the data and permission to transmit it. The LLC layer controls frame synchronization, flow control and error checking.
Physical (Layer 1)	This layer conveys the bit stream-electrical impulse, light or radio signal—through the network at the electrical and mechanical level. It provides the hardware means of sending and receiving data on a carrier, including defining cables, cards and physical aspects.

Words

architecture	n.	建筑，建筑学，体系机构
aspect	n.	样子，外表，面貌，(问题等的)方面
authentication	n.	证明，鉴定
carrier	n.	运送者，邮递员，[电]载波(信号)
compatibility	n.	[计]兼容性，适合，一致(性)，互换性
congestion	n.	拥塞，充血，交通堵塞
constraint	n.	约束，强制，局促
convey	v.	搬运，传达，转让
coordination	n.	同等，调和，配合，(动作)协调
decode	v.	解码，译解
encode	v.	编码，把(电文、情报等)译成电码(或密码)
encrypt	v.	[计]加密，将……译成密码
establish	v.	建立，设立，安置，使定居，确定

framework	n.	构架，框架，结构
furnish	v.	供应，提供
handle	n.	柄，把手
	v.	触摸，处理，操作
hierarchy	n.	层次，层级
host	n.	主机，主人
identify	v.	确定
implement	n.	工具，器具
	v.	贯彻，实现，执行
independence	n.	独立，自主
layer	n.	层，阶层
packet	n.	小包裹，小捆，信息包
	v.	包装
presentation	n.	介绍，陈述，赠送，表达
privacy	n.	独处而不受干扰，秘密
responsible	adj.	有责任的，可靠的，可依赖的，负责的
sequence	n.	次序，顺序，序列
session	n.	会议，开庭
synchronization	n.	同一时刻，同步
syntax	n.	[语]语法，有秩序的排列，句子构造，句法
terminate	v.	停止，结束，终止
tier	n.	列，行，排，层，等级
	v.	使造成递升排列，使层叠

phrases

vice versa		反之亦然
LLC (Logical Link Control)		逻辑链路控制
MAC (Media Access Control)		媒体访问控制
OSI (Open System Interconnection)		开放系统互连

2.1.2 正文参考译文

计算机网络介绍

计算机网络是连接两个或多个计算机的系统，它允许用户快速地交换数据，访问和共享包括设备、应用软件和信息在内的资源。

数据通信系统是通过通信线路将数据从一个地方传送到另外一个地方的电子系统。我们可能经常使用数据通信通过自己的微机将信息发送给使用另外一台机器的朋友。我们所在的公司，其计算机系统可能遍布一座大楼，甚至是全国乃至世界。也就是说，所有的部分(输入和输出单元、处理器和存储设备)都在不同的地方，它们通过通信连接起来。我们也可能使用远程通信线(电话线)接进位于外部数据库的信息，然后将信息传送到自己的微机上进行重新加工和分析。

为了连接到网络上,需要使用特殊用途的硬件部件来处理所有的传送。这个硬件被称为是网络适配卡或网络接口卡,它是插入到计算机总线上的输出电路板,由电缆将它连接到网络设备上。

通信网络根据其占据的地理范围不同,而被分为三种重要的类型:局域网、城市网和广域网。

局域网 计算机和外部设备在很近的物理范围内(例如在一座大楼内)的网络被称为是局域网。局域网由电缆电话线、同轴电缆或光缆连接,通常使用总线型的结构。在局域网中人们可以共享不同的设备,这样可以降低设备的费用。局域网可以通过使用网关连接到另外一个局域网或者更大的网。使用网关,一个局域网可以被连接到另一个办公团体的局域网,也可以被连接到世界范围的其他局域网,即使它们的配置不同。另外,也可以用网桥来连接具有相同配置的网络。

现在有了一种新的局域网:无线局域网。无线局域网是灵活的数据传输系统,实现了大楼或校园内有线局域网的延伸或替换。无线局域网使用电磁波通过空气传送和接收数据,最低限度地减少了有线连接。无线局域网把数据连接和用户移动性结合起来,通过简单的配置,形成了移动的局域网。

随着近几年的发展,无线局域网在一些领域已经获得了普及,其中包括健康保健、零售业、制造业、仓储业和学术界。这些行业通过手提终端和笔记本电脑将实时信息传送到中央主机进行处理,提高生产率,已获益匪浅。如今,对于广泛的商业客户来说,无线局域网正成为公认的通用连接的替代品。

无线局域网的应用 无线局域网常常扩充而并非代替有线局域网的功能,它通常提供骨干网络和移动用户间最后几米的连接。通过无线局域网的灵活性和功能可以实现许多应用,以下描述了其中的一部分:

- 医院的医生和护士利用手提或笔记本电脑与无线局域网连接的性能,及时传递了病人的信息,提高了效率;
- 顾问或会计审计事务组或一些小的工作组使用快速搭建的网络提高了工作效率;
- 动态环境下的网络管理者使用无线局域网最大限度地减少了经常要进行的移动、添加和修改工作,从而降低了局域网所有者的费用;
- 公司的培训点以及大学的学生使用无线连接以便于访问信息、进行信息交换以及学习;
- 在旧的建筑物内无线局域网是经济实惠的网络基础结构的解决方案;
- 零售商店的老板使用无线局域网简化经常性的网络重新配置(问题);
- 贸易展览部门工作人员通过安装预先配置的无线局域网最大限度地降低了配置需求,而不需要当地信息管理系统的支持;
- 仓储工人使用无线局域网和中心数据库交换信息,提高了生产效率;
- 网络管理员使用无线局域网提供运行在有线网络上的关键应用程序的备份;
- 因为手头有实时信息可供使用,在会议室的高级行政官因此可以做出快速的决定。

日益增长的移动用户也成为无线局域网的坚实的后备力量。使用膝上电脑和无线网络接口卡就可实现移动访问无线局域网,这就使得用户在不同的地方(会议室、门厅、休息室、自助食堂、教室等)穿梭时仍然可以访问其网络数据。假如没有无线局域网,用户就不得不携带笨重的电缆寻找网络插头。

城域网 这些网络用于一个城市内的建筑物之间的连接。移动电话系统通过允许将汽车电话和移动电话接入而扩展了城域网的灵活性。

广域网 广域网是国家和世界范围内的网络。在其他的信道种类中，广域网使用微波中继和卫星通信远距离到达用户。使用最广泛的广域网是 Internet，它允许用户和用户及设备在世界范围内进行连接。

2.1.3 阅读材料参考译文

七层开放系统互联模型

开放系统互联模型把实现协议的网络架构分为七层，控制从一层传到下一层，从一个站的应用层开始，依次进行到最底层，通过传输通道到达下一层，再逐层返回。

应用层 (第7层)	应用层支持应用程序和终端用户处理。确定通信双方和服务质量，考虑用户认证和隐私，确定数据语法的约束。这一层是专用于应用的。这一层为文件传输、E-mail 以及其他网络软件服务提供应用服务。Telnet 和 FTP 是完全存在于应用层的应用程序。应用程序体系结构分级是本层的部分
表示层 (第6层)	表示层将数据转换为应用层能接收的格式。该层格式化和加密即将送向网络的数据，而不用考虑兼容性问题。它有时也被称为语法层
会话层 (第5层)	该层建立、管理和终止应用程序之间的连接。在会话层建立、协调和终止每个终端上应用程序之间的会话、交流和对话。它涉及会话和连接
传输层 (第4层)	该层提供在终端系统或主机之间的透明数据传输，并负责端到端的错误恢复和流量控制。它确保完成数据传输
网络层 (第3层)	该层提供交换和路由技术，建立逻辑路径(被称为虚电路)，从节点到节点传输数据。该层具有路由和转发功能，以及寻址、互联网络、错误处理、拥塞控制和分组排序功能
数据链路层 (第2层)	在该层，数据包被编码和解码为二进制位(比特，bit)。它提供了物理层上的传输协议、管理和错误处理，以及流量控制和帧同步。数据链路层分为两个子层：媒体访问控制(MAC)层和逻辑链路控制(LLC)层。MAC 子层控制网络上的计算机获取数据，并允许传送数据。LLC 子层负责帧同步控制、流量控制和错误检查
物理层 (第1层)	该层在电力和机械标准下通过网络传输比特流(电脉冲、光或无线电信号)。它提供发送和接收数据信号的硬件方法，包括确定电缆、卡和物理状况

2.2 Data Communications Channels

To get here to there, data must move through something. A telephone line, cable, or the atmosphere are all transmission media or channels. But before the data can be communicated, it must be converted into a form suitable for communication.

Data communication lines can be connected in two types of configurations: point-to-point and multidrop. A point-to-point line directly connects the sending and the receiving devices, and a multidrop line connects many devices, not just one sending device and one receiving device.

The two ways of connecting microcomputers with each other and with other equipments are through the cable and through the air. There are three basic forms into which data can be converted for communication: electrical pulses or charges, electromagnetic waves, and pulses of light.

Specifically, five kinds of technology are used to transmit data. These are telephone lines (twisted pair), coaxial cable, fiber-optic cable, microwave and satellite.

Telephone Lines [1] Inexpensive, multiple-conductor cable comprised of one or more pairs of 18 to 24 gauge copper strands. The strands are twisted to improve protection against electromagnetic and radio frequency interference. The cable, which may be either shielded or unshielded, is used in low-speed communications, as telephone cable. It is used only in baseband networks because of its narrow bandwidth. Most telephone lines you see strung on poles that consist of cables made up of hundreds of copper wires are twisted pairs. Twisted pairs are susceptible to a variety of types of electrical interference (noise), which limits the practical distance that data can be transmitted without being garbled. Twisted pairs have been used for years for voice and data transmission, however they are now being phased out by more technically advanced and reliable media.

Coaxial Cable Coaxial cable is a type of thickly insulated copper wire that can carry a larger volume of data—about 100 million bits per second, the insulation is composed of a nonconductive material covered by a layer of woven wire mesh and heavy-duty rubber or plastic. In terms of number of telephone connections, a coaxial cable has over 80 times the transmission capacity of twisted pair. Coaxial cables are most often used as the primary communications medium for local connected network in which all computer communication is within a limited geographic area, such as in the same building.

Coaxial cable is also used for undersea telephone lines.

Fiber-Optic Cable [2] A transmission medium composed of a central glass optical fiber cable surrounded by cladding and an outer protective sheath. It transmits digital signals in the form of modulated light from a laser or LED (light-emitting diode). In fiber-optic cable, data is transmitted as pulses of light through tubes of glass. In terms of number of telephone connections, fiber-optic cable has over 20,000 times the transmission capacity of twisted pair. However, it is significantly smaller. Indeed, a fiber-optic tube can be half the diameter of a human hair. Although limited in the distance they can carry information, fiber-optic cables have several advantages. Such cables are immune to electronic interference, which makes them more secure. They are also lighter and less expensive than coaxial cable and are more reliable at transmitting data. They transmit information using beams of light at light speeds instead of pulses of electricity, making them far faster than copper cable. Fiber-optic cable is rapidly replacing twisted-pair telephone lines.

Microwave Instead of using wire or cables, microwave systems can use the atmosphere as the medium through which to transmit signals. Microwaves are high-frequency radio waves that travel in straight lines through the air. Because the waves cannot bend with the curvature of the earth, they can be transmitted only over short distances. Thus, microwave is a good medium for sending data between buildings in a city or on a large college campus. For longer distances, the waves must be relayed by means of "dishes", or antennas. These can be installed on towers, high buildings and mountaintops. Each tower facility receives incoming traffic, boosts the signal strength, and sends the signal to the next station.

Satellites [3]Satellite communications refers to the utilization of geostationary orbiting satellites to relay the transmission received from one earth station to one or more earth stations. They are the outcome of research in the area of communications whose objective is to achieve ever-increasing ranges and capacities with the lowest possible costs. Orbiting about 22,000 miles above the earth, satellites rotate at a precise point and speed above the earth. This makes them appear stationary so they can amplify and relay microwave signals from one transmitter on the ground to another. The primary advantage of satellite communication is the amount of area that can be covered by a single satellite. It also has other features: long communication distance, and the cost of station building is independent of the communication distance, operating in broadcasting mode, easy for multiple access, sustaining heavy traffic, able to transport different types of service, independent sending and receiving, and monitoring. Three satellites placed in particular orbits can cover the entire surface of the earth, with some overlap. Their only drawback is that bad weather can sometimes interrupt the flow of data.

Words

antenna	n.	天线
axis	n.	轴心，中轴
baseband	n.	基带
boost	v.	增加，放大
cable	n.	电缆
charge	n.	电荷
cladding	n.	金属包层，外罩
coaxial	adj.	同轴的，共轴的
configuration	n.	配置，结构，布局，格局
curvature	n.	弯曲，曲度，弧度
electromagnetic	adj.	电磁的
fiber-optic	n.	光纤
garble	v.	精选，筛去……的杂质，误解
gauge	n.	规格，标准

geostationary	*adj.*	与地球相对位置保持不动的
interference	*n.*	干扰，干涉
immune	*adj.*	免疫的，免除的，不受影响的
multidrop	*n.*	多点
multiple-conductor	*n.*	多重导体
outcome	*n.*	结果，产量，出口
pole	*n.*	极(点)，杆，电极
pulse	*n.*	脉冲
relay	*v.*	中继，转播
sheath	*n.*	外皮，外层覆盖物
shield	*n.*	屏，保护，防护，屏蔽
strand	*n.*	绳股，绞合线
susceptible	*adj.*	灵敏的，敏感的
utilization	*n.*	使用，利用
woven	*adj.*	纺织的，编织的

Phrases

coaxial cable	同轴电缆
fiber-optic cable	光缆
geostationary orbiting satellite	轨道通信卫星
phase out	逐步停止采用，逐步退出
twisted-pair	双绞线
woven wire	钢丝网，铁丝网

Abbreviations

LED(Light-Emitting Diode)	发光二极管

Notes

[1] 例句：Inexpensive, multiple-conductor cable comprised of one or more pairs of 18 to 24 gauge copper strands.

分析：这句话实际上是一个名词解释，过去分词短语 comprised of ...作定语，修饰 cable。

译文：由一对或多对18到24规格的铜线组成的并不昂贵的多重导体电缆。

[2] 例句：A transmission medium composed of a central glass optical fiber cable surrounded by cladding and an outer protective sheath.

分析：这句话也是一个名词解释，过去分词短语 composed of...作定语，修饰 medium，同样过去分词短语 surrounded by...作定语，修饰 cable。

译文：由金属包裹层和外层保护层包裹的、中央玻璃光学纤维绳组成的一种传输介质。

[3] 例句：Satellite communications refers to the utilization of geostationary orbiting satellites to relay the transmission received from one earth station to one or more earth stations.

分析：本句为简单句，句中 refer to 意思是"指的是"。例如：Loosely, the term "job" is sometimes used to refer to a representation of a job. 习惯上说，"作业"这个术语通常指的是作业的一种表述。

译文：卫星通信指的是利用轨道通信卫星把地面站发来的信息传送到另外一个或多个地面站。

Exercises

Ⅰ. Put "true" or "false" in the brackets for the following statements according to the passage.

1. (　) Air can be used to transmit data.
2. (　) Point-to-point and multidrop are two types of configuration for communications lines connecting.
3. (　) Twisted pairs are susceptible to noise, which limits the practical distance they can transmit.
4. (　) Coaxial cable and twisted pairs are all high-frequency transmission cables.
5. (　) Optic-fiber cables can transmit both electrical and light signals.
6. (　) Optic-fiber cables are more secure because they are not susceptible to electrical interference.
7. (　) Microwave can transmit data over long distances.
8. (　) The primary advantage of satellite communication is the long distance a single satellite can cover.
9. (　) Three satellites placed in random orbits can cover the entire surface of the earth.
10. (　) All the five kinds of communications channels have their own advantages and disadvantages, so they can not replace with each other.

Ⅱ. Fill in the blanks according to the passage.

1. The two ways of connecting microcomputers with each other and with other equipments are through the _____ and through the ____.
2. Data communications lines can be connected in two types of configurations: _____ and _____.
3. Twisted pairs are _____ to a variety of types of electrical interference (noise), which limits the practical distance that data can be transmitted without being _____.
4. _____ transmits digital signals in the form of modulated light from a laser or LED
5. _____ systems can use the atmosphere as the medium through which to transmit signals.
6. Microwaves cannot bend with the _____ of the earth, they can be transmitted only over short distances.
7. Satellite communications refers to the _____ of geostationary orbiting satellites to

_____ the transmission received from one earth station to one or more earth stations.

8. Three satellites placed in particular orbits can cover the entire surface of the earth, with some _____.

9. Fiber-optic cables are _____ to electronic interference, which makes them more secure.

10. In order to transmit longer distance, microwave must be relayed by means of _____.

III. Translate the following words and expressions into Chinese.

1. communications channels
2. multiple-conductor
3. facility
4. stationary
5. transmission capacity
6. electromagnetic waves
7. pulses of light
8. light-emitting diode
9. a solid wire conductor
10. electrical interference

2.2.1 Reading Material

Computer networking device

Computer networking devices are units that mediate data in a computer network. Units which are the last receiver or generate data are called hosts or data terminal equipment.

List of computer networking devices

Common basic network devices:

- Gateway: Device sitting at a network node for interfacing with another network that uses different protocols. Works on OSI layers 4 to 7.
- Router: A specialized network device that determines the next network point to which to forward a data packet toward its destination. Unlike a gateway, it cannot interface different protocols. Works on OSI layer 3.
- Bridge: A device that connects multiple network segments along the data link layer. Works on OSI layer 2.
- Switch: A device that allocates traffic from one network segment to certain lines (intended destination(s)), which connect the segment to another network segment. So unlike a hub a switch splits the network traffic and sends it to different destinations rather than to all systems on the network. Works on OSI layer 2.
- Hub: Connects multiple Ethernet segments together making them act as a single segment. When using a hub, every attached device shares the same broadcast domain and the same collision domain. Therefore, only one computer connected to the hub is able to transmit at a time. Depending on the network topology, the hub provides a basic level 1 OSI model connection among the network objects (workstations, servers, etc). It provides bandwidth that is shared among all the objects, compared to switches, which provide a dedicated connection between individual nodes. Works on OSI layer 1.
- Repeater: Device to amplify or regenerate digital signals received while setting them from one part of a network into another. Works on OSI layer 1.

Hardware or software components that typically sit on the connection point of different networks, e.g. between an internal network and an external network:
- Proxy: computer network service which allows clients to make indirect network connections to other network services.
- Firewall: a piece of hardware or software put on the network to prevent some communications forbidden by the network policy.
- Network Address Translator: network service provide as hardware or software that converts internal to external network addresses and vice versa.

Other hardware for establishing networks or dial-up connections:
- Multiplexer: Device that combines several electrical signals into a single signal.
- Network Card: A piece of computer hardware to allow the attached computer to communicate by network.
- Modem: Device that modulates an analog "carrier" signal (such as sound), to encode digital information, and that also demodulates such a carrier signal to decode the transmitted information, as a computer communicating with another computer over the telephone network.
- ISDN terminal adapter (TA): A specialized gateway for ISDN.
- Line Driver: A device to increase transmission distance by amplifying the signal. Base-band networks only.

Words

amplify	v.	放大,增强,扩大
bandwidth	n.	[电信]带宽,频带宽度
bridge	n.	桥
convert	v.	使转变,转换……,使……改变信仰
dedicated	adj.	专注的,献身的
demodulate	v.	[讯]使解调,使检波
destination	n.	目的地,终点,[计]目的文件,目的单元格
determine	v.	决定,确定,测定,使下定决心
firewall	n.	防火墙
forbidden	adj.	禁止的,严禁的
hub	n.	网络集线器,网络中心
hybrid	n.	混合物
	adj.	混合的
interface	n.	[地质]分界面,接触面,[物、化]界面
internal	adj.	内在的,国内的
interwork	v.	互相作用,互通
mediate	v.	仲裁,调停,作为引起……的媒介
modem	n.	调制解调器

multiplexer	n.	多路(复用)器
packet	n.	小包裹，小捆，信息包
proxy	n.	代理人，代理服务器，即 proxy 服务器
regenerate	v.	使新生，重建，再生
	adj.	新生的，更新的
router	n.	[计] 路由器
split	v.	劈开，(使)裂开，分裂，分离
	n.	裂开，裂口，裂痕
switch	n.	交换机，交换器，开关，电闸，转换
traffic	n.	交通，运输，交通量，通信量
	v.	交易，买卖

Phrases

broadcast domain	广播域
collision domain	碰撞域
data terminal equipment	数据终端设备
ISDN terminal adapter	综合业务数字网终端适配器
Line Driver	线路驱动器
Network Address Translator	网络地址转换器
Network Card	网卡
Network Device Connectivity	网络设备连接

2.2.2 正文参考译文

数据通信信道

为了传递数据，必须有一些介质，电话线、电缆或空气都是传输介质，即信道。但是，数据在传送前必须先被转化成适于通信的形式。

数据通信线路可以以两种配置方式连接：点到点和多点线路连接。点到点线路直接连接发送设备和接收设备，多点线路连接多个设备而不仅一个发送设备和一个接收设备。

微型机之间以及与其他设备的连接一般通过电缆和通过空气两种方法。数据可以被转化为用于通信的三种基本形式：电脉冲或电荷、电磁波和光波。

具体来说，有五种传送数据的方式。它们是电话线(双绞线)、同轴电缆、光缆、微波和卫星。

电话线 即由一对或多对 18 到 24 规格的铜线组成的并不昂贵的多重导体电缆。铜线绞合着以提高防止电磁和无线电频率干扰的能力。这些电缆，可以是屏蔽的也可以是非屏蔽的，作为电话线用于低速通信中。由于带宽窄，因此仅被用于基带网络中。架在电线杆上的绝大多数电话线是双绞线，它们是由数以百计的铜线组成的电缆。双绞线对各类电的干扰很敏感，因此它传送的数据很容易被窃改。双绞线被用于传递声音和数据已有好多年了，然而它们正被技术更先进和可靠的介质逐步代替。

同轴电缆 同轴电缆是一种有厚绝缘层的铜线,可以携带大量的数据——大约每秒100

万位。这个绝缘层由覆盖有一层金属筛网的绝缘材料和厚的橡胶或塑料组成。根据电话连接的数目，同轴电缆的传输容量可达双绞线的80多倍。同轴电缆是局域网最常用的且很主要的通信介质，而局域网中所有计算机间的通信是局限在有限的地理区域(比如一座大楼内)。

同轴电缆也可用作海底电话线。

光缆 它是由金属包裹层和外层保护层包着的、中央玻璃光学纤维绳组成的一种传输介质。光缆以调制的从激光或光发射二板管中发射的光的形式来传递数字信号。在光缆中，当光脉冲通过玻璃管时数据就被传送。由于可连接的电话很多，光缆的传输容量可达双绞线的2万多倍，然而它的体积却非常小。实际上，光缆管的直径仅仅是人的头发丝的一半。虽然传送信息的距离有限，但它具有一些优点：不受电子的干扰，这就使得它们更安全；它们比同轴电缆轻、便宜，而且传输数据更可靠；光缆用光束以光的速度而不是电脉冲的速度传输数据，这就使得通过光缆传输比通过铜缆传输快得多。因此光缆正迅速地取代双绞线电话线。

微波 它不使用电线或电缆，而使用大气作为传输介质传送信号。微波是以直线穿过空气的高频率的无线电波。由于微波不能随地球表面的弯曲度弯曲，所以只能传送很短的距离。这样，对于一个城市中或一个大学校园内的两个大楼之间，微波是很好的传送数据的介质。对于较长距离，微波必须通过"盘子"(即天线)放大。这些天线可以安装在塔上、高楼上以及山顶上。它们接收信号，增强信号的强度，然后将信号发送给下一站。

卫星 卫星通信指的是利用轨道通信卫星把地面站发来的信息传送到另外一个或多个地面站。卫星通信是通信领域的研究成果，其目的就是以尽可能少的代价获得不断增加的信息范围和容量。卫星在距地球表面22 000英里的轨道上在精确的位置以精确的速度旋转着。这就使得它们看起来是静止的，从而可以将一个地面站转发器发送来的信息放大并传送到下一站。通信卫星的主要优点是一个卫星可以覆盖的面积很大。它还具有其他的特点：远距离通信，转发站的费用独立于通信距离，以广播方式运作，易于多路访问，承受大量的信息，能够传送不同类型的服务，独立地发送、接收和监控。在特定轨道放置三颗卫星就可覆盖整个地球表面，而且还有些重叠。其惟一缺点是恶劣的天气有时会影响数据的传送。

2.2.3 阅读材料参考译文

计算机网络设备

计算机网络设备在计算机网络中负责传输数据。最终接收或产生数据的设备称为主机或数据终端设备。

计算机网络设备列表

常见的基本网络设备：

- 网关：在网络节点与另一个使用不同协议的网络进行连接的装置。工作在OSI模型的第4至第7层。
- 路由器：一个专门的网络设备，决定传输数据包到目的地的下一个网络点。不同于网关，它不能连接不同的协议。工作在OSI模型的第3层。
- 网桥：连接数据链路层沿线的多个网段的装置。工作在OSI模型的第2层。

- 交换机：一个分配流量从一个网段到某些线路(预期的目的地)的设备，这些线路将网段连接到另一个网段。因此，不同于集线器，一个交换机分解网络流量并传送到不同的目的地，而不是传送到网络上的所有系统。工作在 OSI 模型的第 2 层。
- 集线器：把多个以太网段连接在一起，使它们作为一个单一的网段。当使用集线器时，每个连接的设备共享相同的广播域和相同的碰撞域。因此，某个时间只有一台连接到集线器的计算机能够传输。利用网络的拓扑结构，集线器在网络对象(工作站、服务器等)之间提供了一个基本的 OSI 模型第 1 层的连接。它提供的带宽被所有对象共享，与交换机不同，集线器提供各节点之间的专用连接。工作在 OSI 模型的第 1 层。
- 中继器：该设备在把数字信号从网络的一部分传到另一部分时，放大或再生接收的数字信号。工作在 OSI 模型的第 1 层。

在不同网络的连接节点(例如在内部网络和外部网络之间)上的硬件或软件组件：
- 代理：让客户能够间接连接到其他网络服务的计算机网络服务。
- 防火墙：放在网络上的硬件或软件，以阻止网络政策所禁止的一些通信。
- 网络地址转换：硬件或软件提供的将内部网络地址转换为外部网络地址(反之亦然)的网络服务。

建立网络或拨号连接的其他硬件：
- 复用器：将多路电信号结合成为一个单路电信号的设备。
- 网卡：使连接的计算机能够通过网络进行通信的一块计算机硬件。
- 调制解调器：该设备将模拟"载波"信号(如声音)调制为已编码的数字信息，而且还可以解调这种载波信号对传输信息的解码，以使一台计算机通过电话网与另一台计算机通信。
- 综合业务数字网终端适配器(TA)：专门用于 ISDN 的网关。
- 线路驱动器：它通过放大信号增加传输距离，且仅用于基带网络。

2.3 Main Factors Affecting Data Transmission

There are several factors that affect data transmission. They include speed or bandwidth, serial or parallel transmission, direction of data flow, modes of transmission data, and protocols.

Bandwidth The different communications channels have different data transmission speeds. This bit-per-second transmission capability of a channel is called its bandwidth. Bandwidth may be of three types: voiceband, medium band and broadband. Voiceband is the bandwidth of a standard telephone line and used often for microcomputer transmission, the bps is 300~9600. Medium band is the bandwidth of special liased lines used mainly with minicomputers and mainframe computers, the bps is 56,000~264 million. Broadband is the bandwidth that includes microwave, satellite, coaxial cable, and fiber-optic channels. It is used for very high-speed computers whose processors communicate directly with each other. It is in the range of 56,000~30 billion bps.

Serial or Parallel Transmission Data travels in two ways: serially and in parallel. In serial data transmission, bits flow in a serial or continuous stream, like cars crossing a one-lane bridge. Each bit travels on its own communications line. [1]Serial transmission is the way most data is sent over telephones lines. Thus, the plug-in board making up the serial connector in a microcomputer's modem is usually called a serial port. More technical names for the serial port are RS-232C connector and asynchronous communications port. With parallel data transmission, bits flow through separate lines simultaneously. In other words, they resemble cars moving together at the same speed on a multilane freeway. Parallel transmission is typically limited to communications over short distances and is not used over telephone lines. It is, however, a standard methods of sending data from a computer's CPU to a printer.

Direction of Data Transmission There are three directions or modes of data flow in a data communications system: simplex communication, half-duplex communication, and full-duplex communication. Simplex communication resembles the movement of cars on a one-way street. Data travels in one direction only. It is not frequently used in data communication systems today. One instance in which it is used may be in point-of sale (POS) terminals in which data is being entered only. In half-duplex communication, data flows in both directions, but not simultaneously. That is, data flows in only one direction at any one time. This resembles traffic on a one-lane bridge. Half-duplex is very common and is frequently used for linking microcomputers by telephone lines to other microcomputers, minicomputers, and mainframes. Thus, when you dial into an electronic bulletin board through your microcomputer, you may well be using half-duplex communication. In full-duplex communication, data is transmitted back and forth at the same time, like traffic on a two-way street. It is clearly the fastest and most efficient form of two-way communication. However, it requires special equipment and is used primarily for mainframe communications. An example is the weekly sales figures that a supermarket or regional office sends to its corporate headquarters in another place.

Modes of Transmitting Data Data may be sent by asynchronous or synchronous transmission. In asynchronous transmission, the method frequently used with microcomputers, data is sent and received one byte a time. Asynchronous transmission is often used for terminals with slow speeds. Its advantage is that the data can be transmitted whenever convenient for the sender. Its disadvantage is a relatively slow rate of data transfer. Synchronous transmission is used to transfer great quantities of information by sending several bytes or a block at a time. For the data transmission to occur, the sending and receiving of the blocks of bytes must occur at carefully timed intervals. Thus, the system requires a synchronized clock. Its advantage is that data can be sent very quickly. Its disadvantage is the cost of the required equipment.

Protocols For data transmission to be successful, sender and receiver must follows a set of communication rules for the exchange of information. These rules for exchanging data between computers are known as the line protocol. A communication software package like Crosstalk helps define the protocol, such as speeds and modes, for connecting with another microcomputer. TCP/IP (Transmission Control Protocol and Internet Protocol) are the two standard protocols for communications on the Internet.

TCP/IP is the "language" of the Internet. It is a networking technology developed by the United States Government Defense Advanced Research Project Agency (DARPA) in the 1970s. It is most commonly employed to provide access to the Internet but can be and is used by many people to create a LAN that may or may not connect to the Internet. In many aspects TCP/IP is a client/server-type LAN, but many manufacturers of TCP/IP software have applications that allow the "clients" to serve files or even applications. TCP/IP is truly an open systems protocol. This means that no one manufacturer creates the product—any computer running TCP/IP software can connect to anyone else who has TCP/IP software (provided the user has an account and security permissions), regardless of who made the particular version of software.

When different types of microcomputers are connected in a network, the protocols can become very complex. Obviously, for the connections to work, these network protocols must adhere to certain standards. The first commercially available set of standards was IBM's Systems Network Architecture (SNA). This works for IBM's own equipment, but other machines won't necessarily communicate with them. The International Standards Organization has defined a set of communications protocols called the Open Systems Interconnection (OSI). The purpose of the OSI model is to identify functions provided by any network. [2]It separates each network's functions into seven "layers" of protocols, or communication rules. When two network systems communicate, their corresponding layers may exchange data. [3]This assumes that the microcomputers and other equipment on each network have implemented the same functions and interfaces.

Words

adhere	v.	黏着于(坚持，追随)……
assume	v.	假定，设想，采取，呈现
asynchronous	adj.	异步的
broadband	n.	宽频带，宽波段
bulletin	n.	公报，通报，告示
duplex	adj.	双工的，双向的
implement	v.	实现，完成
interface	n.	界面
parallel	adj.	平行的，并联的
protocol	n.	协议
serial	adj.	连续的，串联的
simplex	adj.	单工的，单向的
voiceband	n.	话音频带

Phrases

dial into	拨入
open system protocol	开放系统协议

Abbreviations

OSI(Open Systems Interconnection)　　　开放系统互连
POS(Point-Of-Sale)　　　　　　　　　　零售点
SNA(Systems Network Architecture)　　　系统网络结构
TCP/IP(Transmission Control Protocol　　传输控制协议和因特网协议
　　　and Internet Protocol)

Notes

[1] 例句：Serial transmission is the way most data is sent over telephones lines.
　　分析：most data is sent over telephones lines 是定语从句，修饰 the way。
　　译文：串行传输是绝大多数数据通过电话线传输的方式。

[2] 例句：It separates each network's functions into seven "layers" of protocols, or communication rules.
　　分析：短语 separate…into…意思是"把……分成……"。
　　译文：它把每个网络的功能分成七层协议，即七层通信规则。

[3] 例句：This assumes that the microcomputers and other equipment on each network have implemented the same functions and interfaces.
　　分析：句中 that 引导的是宾语从句，作 assumes 的宾语。
　　译文：这种情况假设每一个网络中的微机和其他设备实现了同样的功能，具有同样的接口。

Exercises

Ⅰ. Put "true" or "false" in the brackets for the following statements according to the passage.

1. (   ) Protocols is the main factor that affects data communication.

2. (   ) Broadband is suitable for all kinds of communication channels and is mainly used for high-speed computers.

3. (   ) Serial and parallel transmission are the two ways of data traveling, they can't be replaced with each other.

4. (   ) In asynchronous transmission, the method frequently used with microcomputers, data is sent and received several bytes a time.

5. (   ) Simplex communication is very common and is frequently used for linking microcomputers by telephone lines to other microcomputers, minicomputers, and mainframes.

6. (   ) TCP/IP is the "language" of the Internet, which can only be used on internet.

7. (   ) Protocols are a set of communication rules for sender and receiver to exchange data.

8. (   ) An open system protocol means that no special product will be needed, and only those that obey the protocol will be admitted and can share resources.

9. (   ) The purpose of the OSI model is to identify functions provided by any network.

10. (   ) OSI model assumes that the microcomputers and other equipment on each network

have implemented the same functions and interfaces.

II. Fill in the blanks according to the passage.

1. There are several factors that affect data transmission. They include speed or _____, serial or _____ transmission, _____ of data flow, modes of transmission data, and _____.

2. In serial data transmission, bits flow in a serial or _____ stream.

3. With parallel data transmission, bits flow through _____ lines simultaneously.

4. _____ is very common and is frequently used for linking microcomputers by telephone lines to other microcomputers, minicomputers, and mainframes.

5. In _____ communication, data is transmitted back and forth at the same time.

6. Asynchronous transmission is often used for terminals with _____.

7. Synchronous transmission requires a _____ clock.

8. TCP/IP (Transmission Control Protocol and Internet Protocol) are the two standard protocols for communications on the _____.

9. The OSI model separates each network's functions into seven "layers" of _____, or communication rules.

10. TCP/IP is truly an open systems protocol. This means that no one manufacturer creates the product—any computer running TCP/IP software can _____ to anyone else who has _____ software regardless of who made the particular version of software.

III. Translate the following words and expressions into Chinese.

1. protocol
2. modes of transmission data
3. transmission control protocol
4. asynchronous
5. serial transmission
6. full-duplex communication
7. point-of sale terminal
8. broadband
9. electronic bulletin board
10. channel

2.3.1 Reading Material

How to install TCP/IP on your computer

This document discusses checking to see if your computer has TCP/IP installed. Important: If your company has an network, or computer, administrator, then please consult them before making changes to a computer's network settings! The installation of TCP/IP can vary from one version of Windows to another, so make sure you match the instruction set below to your copy of Windows.

Windows XP automatically detects whether your machine is using TCP/IP, and disables it if it is not! So first you must determine whether the machine has TCP/IP installed on a modem or on a network connection or, of course, not at all. In order to do this, please follow these simple instructions:

1) Select the Start button, then the Settings menu, and then the Control Panel from the Settings menu.

2) Double click on Network Connections.

3) This should show a window of all network related connections.

i. If there are no icons under Dial up or LAN, then you don't have any network or dial up connections, and therefore, no TCP/IP. For now, you can close all windows to return to your desktop.

ii. If you don't have any LAN type connections, you can close all windows to return to your desktop. Otherwise, continue with step 4.

4) Right click on the any of the network (those that say LAN) icons and choose Properties.

5) On the General tab, look through the list and see if you can see Internet Protocol TCP/IP. You will need to repeat steps 4 and 5 for all LAN connections in the window.

6) If it is there, you have TCP/IP installed. If not, you don't have TCP/IP.

A) You don't have any network connections, or you have a modem, but need to ensure TCP/IP is enabled: You can install the Microsoft Loopback Adapter to take care of TCP/IP. Follow these easy instructions to install:

1) Select the Start button, then the Settings menu, and then the Control Panel from the Settings menu.

2) Double click on Add Hardware.

3) Click Next on the Add Hardware wizard.

4) Click Yes for Have you got the hardware connected, and then click Next.

5) Select Add new Hardware device from the list, and then Next.

6) Select Install the Hardware device that I manually select from a list (Advanced), and then click Next.

7) Click Network Adapters, and then click Next.

8) In the Manufacturers box, click Microsoft.

9) In the Network Adapters box, click Microsoft Loopback Adapter, and then click Next.

10) Click Next to start the installation.

B) You have a network card installed, but no TCP/IP: You may be using a different protocol on your network card, so TCP/IP may not be installed. Follow these steps to install:

1) Select the Start button, then the Settings menu, and then the Control Panel from the Settings menu.

2) Double click Network and Dial Up Connections.

3) Right click on the LAN connection, and choose Properties.

4) On the general tab, ensure that TCP/IP is not already in the list, and then click Install.

5) Select Protocol form the list, and click Add.

6) Wait a few seconds, and then select Internet Protocol TCP/IP, and click OK.

7) Wait a few seconds, and then check that TCP/IP shows up in the list on the General tab of the Network Connection Properties, and then close all windows to return to the desktop.

Words

adapter	n.	适配器，改编者
automatically	adv.	自动地，机械地
check	v.	检查，核对，寄存，托运
consult	v.	商量，商议，请教；参考，考虑
desktop	n.	[计]桌面；桌上型电脑
disable	v.	使残废，使失去能力
ensure	v.	保证，担保，使安全，确保
icon	n.	图标，肖像
manually	adv.	用手地；手工地
manufacturer	n.	制造业者，厂商
property	n.	财产，所有物，属性，特性
tab	n.	制表，标号，TAB 键，标签，制表符
vary	v.	改变，变更，使多样化，变化，不同

Phrases

Control Panel   [计]控制面板

2.3.2 正文参考译文

影响数据传送的主要因素

影响数据传送的几个因素，包括速度或带宽、串行或并行传送、数据流的方向、数据传送的方式及协议。

带宽　不同的通信信道有不同的数据传送速度。信道每秒钟能传送二进制位的能力被称为带宽。带宽有三种类型：语音频带、中速带宽和宽带。语音频带是标准电话线的带宽，通常被用于微型机的传送，其每秒钟传送的二进制位数是 300 到 9 600。中速带宽主要是由小型机和大型机使用的特殊连接线的带宽，其每秒钟传送的位数是 56 000 到 26.4 亿。宽带包括微波、卫星、同轴电缆和光缆的信道的带宽。它主要用在处理器之间直接进行通信的高速计算机中，其每秒钟传送的位数为 56 000 到 300 亿位。

串行或并行传输　数据以串行和并行两种方式传送。在串行数据传输中，数据位的流动是一串或连续的流量，就像汽车通过只有一个车道的桥。每一位在自己的通信线路上流动。串行传输是绝大多数数据通过电话线传输的方式。因此，构成微机调制解调器的串行连接器的插入板通常被称为串行口。串行口的技术名称是 RS-232C 连接器和异步通信端口。采用并行数据传输时，数据位是通过分离的线路同时流动的。换句话说，它们类似于汽车一起以同样的速度通过多道快车道。并行传送主要局限于短距离的传送，并不通过电话线。它是从计算机的 CPU 到打印机传送数据的标准方式。

数据传输的方向　在数据通信系统中数据流的方向(或方式)有三种：单工通信、半双工通信和全双工通信。单工通信类似于汽车在单向的街上行驶，数据仅以一个方向传送。它并不经常用于现在的数据通信系统中。使用单工通信的一个例子是只输入数据的零售点终端。在半双工通信中，数据以两个方向流动，但不是同时。也就是说，数据在任何时候都只以一个方向流动，这类似于只有一个车道的桥上的交通。半双工通信很普遍，经常应用于通过电话线连接的微型机之间、微型机和小型机及大型机之间的通信中。因此，当通过微机拨入到电子公告牌，也许使用的就是半双工通信。在全双工通信中，数据同时来回地传送，类似于双向街道上的交通。很显然全双工通信是双向通信中最快最有效的形式。但是，全双工通信需要特殊的设备并且主要用于大型机的通信。其应用的一个典型例子是超市或区域办公室每周将其销售的数据发送给在另一个地方的公司总部。

数据传送方式 数据能以异步或同步方式传送。在异步传送方式下，一次只能发送和接收一个字节，是微型机经常使用的方法。异步传送也常用于速度较慢的终端设备。其优点是在适当的时候发送方就可以传送数据。缺点是数据传送速率相对较慢。同步传送一次发送几个字节或数据块，用于大量信息的传送。为了实现数据传送，发送和接收的字节块必须以精确的时间间隔出现。因此，系统需要一个同步时钟。其优点是可以很快地传送数据，缺点是需要一定的设备费用。

协议 为了成功地传送数据，发送方和接收方必须遵循用于信息交换的一套通信规则。这些用于计算机之间交换数据的规则被称为是线路协议。为了和其他微机连接，像Crosstalk这样的通信软件包就帮助定义协议，诸如速度和方式。TCP/IP(传输控制协议和因特网协议)是Internet上用于通信的两个标准协议。

TCP/IP是Internet的"语言"，它是由美国政府防御高级研究项目机构在20世纪70年代开发的网络技术。它最常用于提供访问Internet，但也被许多人用于建立与Internet相联或不联的局域网。在许多方面，TCP/IP是客户服务器类型的局域网，但许多TCP/IP软件的生产商具有允许"客户机"提供文件甚至应用的应用软件。TCP/IP是一个真正的开放系统协议。这就意味着没有一个生产商生产这个产品——任何运行TCP/IP软件的计算机都可以和另一个具有TCP/IP软件的任一机器相联(假设该用户具有账号和安全许可)，而不管是谁制定的这个软件版本。

当不同类型的微机在网络中相联时，协议将变得非常复杂。显然，为了实现通信，这些网络协议必须坚持一定的标准。第一套商业可用标准是IBM公司的系统网络结构，该标准只用于IBM自己的设备，不过，其他的机器不需要与它们交流。国际标准化组织已经制定了一套被称为开放系统互连的通信协议。开放系统互联模式的目的是判断由任一网络提供的功能。它把每个网络的功能分成七层协议，即七层通信规则。当两个网络系统通信时，它们的对应层可以交换数据。这种情况假设每一个网络中的微机和其他设备实现了同样的功能具有同样的接口。

2.3.3 阅读材料参考译文

如何安装TCP/IP协议

本文讨论检查您的电脑是否已安装TCP/IP。重要提示：如果您的公司有网络或计算机的管理员，在改变计算机的网络设置之前请先咨询他们！根据Windows版本不同安装TCP/IP协议应也不同，因此先确认您的Windows版本是否与下述指令集匹配。

Windows XP会自动检测您的计算机是否正在使用TCP/IP，如果不是则停用它！因此，首先必须确定调制解调器或网络连接是否安装了TCP/IP。或者，什么都没有。为了做到这一点,请按照下面的简单说明操作。

  1) 选择**开始**按钮,再选择**设置**菜单,然后从**设置**菜单选择**控制面板**。
  2) 双击**网络连接**。
  3) 这样会显示一个所有与网络相关的连接的窗口。

  i. 如果没有拨号连接或局域网图标,说明您的电脑没有任何网络或拨号连接,因此,没有 TCP/IP 协议。这时应先关闭所有窗口返回到桌面。

  ii. 如果没有任何局域网连接类型,可以关闭所有窗口返回到桌面。否则,请继续步骤4)。

  4) 右击网络(即局域网)图标,并选择属性。
  5) 在**常规**选项卡中,查看列表,看看是否可以看到互联网协议 TCP/IP。您需要对窗口中的所有 LAN 连接重复步骤 4)和 5)。
  6) 如果列表中有,表明已经安装了 TCP/IP。如果没有,则说明没有安装 TCP/IP。

  A) 您没有任何网络连接,或您有调制解调器,但需要确保 TCP/IP 协议已启用,这时可以安装 Microsoft 环回适配器以处理 TCP/IP 协议。按照以下的简单说明安装:

  1) 选择**开始**按钮,再选择**设置**菜单,然后从**设置**菜单选择**控制面板**。
  2) 双击**添加硬件**图标。
  3) 单击添加硬件向导中的**下一步**按钮。
  4) 在**我已经连接了此硬件**项单击"是"按钮,然后单击**下一步**。
  5) 从列表里选择**添加新的硬件设备**,然后单击**下一步**。
  6) 选择**安装我手动从列表选择的硬件(高级)**,然后单击**下一步**。
  7) 单击**网络适配器**,然后单击**下一步**。
  8) 在制造商框中单击 **Microsoft**。
  9) 在网络适配器框中单击 **Microsoft 环回适配器**,然后单击**下一步**。
  10) 单击**下一步**开始安装。

  B) 您已经安装了一个网络卡,但没有 TCP/IP 协议:您可能在网卡上使用了不同协议,所以可能尚未安装 TCP/IP 协议,这时按照以下步骤安装:

  1) 选择**开始**按钮,再选择**设置**菜单,然后从**设置**菜单选择**控制面板**。
  2) 双击**网络和拨号连接**。
  3) 右击**局域网连接**,并选择**属性**。
  4) 在**常规**选项卡上确认 TCP/IP 不在清单中,然后单击**安装**。
  5) 从列表中选择协议,并单击**添加**。
  6) 等待几秒钟,选择 **Internet 协议 TCP/IP**,然后单击**确定**。
  7) 等待几秒钟,确保 TCP/IP 出现在网络连接属性的**常规**选项卡列表中,然后关闭所有窗口回到桌面。

2.4　Network Architecture

Network architecture describes how computer network is arranged and how computer resources are shared.

There are a number of specialized terms that describe computer network. Some terms often used with networks are: node, client, server, network operating system, distributed processing and host computer.

A node is any device that is connected to a network. It could be a computer, printer, or communication or data storage device.

A client is a node that requests and uses resources available from other nodes. Typically, a client is a user's microcomputer.

A server is a node that shares resources with other nodes. Depending on the resources shared, it may be called a file server, printer server, communication server, or database server.

Network operating system likes Windows, it controls and coordinates the activities between computers on a network. These activities include electronic communication, information, and resource sharing.

In a distributed processing system, computing power is located and shared at different locations. [1]This type of system is common in decentralized organizations where divisional offices have their own computer systems. The computer systems in the divisional offices are networked to the organization's main or centralized computer.

Host computer is a large centralized computer, usually a minicomputer or a mainframe.

A network may consist only of microcomputers, or it may integrate microcomputers or other devices with large computers. [2]Networks can be controlled by all nodes working together equally or by specialized nodes coordinating and supplying all resources. Networks may be simple or complex, self-contained or dispersed over a large geographical area.

Configuration   A network can be arranged or configured in several different ways. The four principal configurations are star, bus, ring, and hierarchical.

In a star network, a number of small computers or peripheral devices are linked to a central unit. This central unit may be a host computer or a file server. All communications pass through this central unit. Control is maintained by polling. That is, each connecting device is asked whether it has a message to send. Each device is then in turn allowed to send its message. One particular advantage of the star form of network is that it can be used to provide a time-sharing system. That is, several users can share resources ("time") on a central computer. The star is a common arrangement for linking several microcomputers to a mainframe that allows access to an organization's database.

In a bus network, each device in the network handles its own communications control. There is no host computer. All communications travel along a common connecting cable called a bus. As the information passes along the bus, it's examined by each device to see if the information is intended for it. The bus network is typically used when only a few microcomputers are to be linked together. This arrangement is common in systems for electronic mail or for sharing data stored on different microcomputers. The bus network is not as efficient as the star network for sharing common resources. (This is because the bus network is not a direct link to the resource.) However, a bus network is less expensive and is in very common use.

In a ring network, each device is connected to two other devices, forming a ring. There is no central file server or computer. Messages are passed around the ring until they reach the correct destination. With microcomputers, the ring arrangement is the least frequently used of the four networks. However, it is often used to link mainframes, especially over wide geographical areas. These mainframes tend to operate fairly autonomously. They perform most or all of their own processing and only occasionally share data and programs with other mainframes. A ring network is useful in a decentralized organization because it makes possible a distributed data processing system. That is, computers can perform processing tasks at their own dispersed locations. However, they can also share programs, data and other resources with each other.

The hierarchical network consists of several computers linked to a central host computer, just like a star network. However, these other computers are also hosts to other, smaller computers or to peripheral devices. Thus, the host at the top of the hierarchy could be a mainframe. The computers below the mainframe could be minicomputers, and those below, microcomputers. The hierarchical network—also called a hybrid network—allows various computers to share databases, processing power, and different output devices. A hierarchical network is useful in centralized organizations. For example, different departments within an organization may have individual microcomputers connected to departmental minicomputers. The minicomputers in turn may be connected to the corporation's mainframe, which contains data and programs accessible to all.

Strategies Every network has a strategy or way of coordinating the sharing of information and resources. The most common network strategies are peer-to-peer and client/server systems.

In a peer-to-peer network system nodes can act as both servers and clients. For example, one microcomputer can obtain files located on another microcomputer and can also provide files to other microcomputers. A typical configuration for a peer-to-peer system is the bus network. Commonly used net operating systems are Apple's Macintosh Peer-to-Peer LANs, Novell's Netware Lite, and Microsoft's Windows for Workgroups. There are several advantages to using this type of strategy. The networks are inexpensive and easy to install, and they usually work well for smaller systems with less than ten nodes. As the number of nodes increases, however, the performance of the network declines. Another disadvantage is the lack of powerful management software to effectively monitor a large network's activities. For these reasons, peer-to-peer network are typically used by small networks.

Client/server network systems use one powerful computer to coordinate and supply services to all other nodes on the network. This strategy is based on specialization. Server nodes coordinate and supply specialized services, and client nodes request the services. Commonly used net operating systems are Novell's Netware, Microsoft's LAN and Windows NT. One advantage of client/server network systems is their ability to handle very large networks efficiently. Another advantage is the powerful network management software that monitors and controls the network's activities. The major disadvantages are the cost of installation and maintenance.

Words

autonomously	adv.	自主地
centralized	adj.	集中的，中央集权的
destination	n.	目的地，目标
decentralized	adj.	分散(型)的
dispersed	adj.	分散的，漫布的，细分的
divisional	adj.	分开的
evolve	v.	进化，开展
hierarchical	adj.	分层的
hose	n.	软管
	v.	接以软管
integrate	v.	使……结合，使……完整
maintenance	n.	维护，保持，保养
node	n.	节点，分支
polling	n.	轮询，探询，查询
self-contained	adj.	独立的，配套的
specialization	n.	专门，专业化
strategy	n.	战略

Phrases

distributed processing system	分布式处理系统
file server	文件服务器
host computer	主机
in turn	轮流，依次
peer-to-peer	对等，对等网络
resource sharing	资源共享
time sharing	分时

Notes

[1] 例句：This type of system is common in decentralized organizations where divisional offices have their own computer systems。

分析：句中 where 引导的是定语从句，用来修饰 organizations。

译文：这类系统在分散型机构中很常见，其分开的办公室具有它们自己的计算机系统。

[2] 例句：Networks can be controlled by all nodes working together equally or by specialized nodes coordinating and supplying all resources.

分析：分词短语 working together equally 作定语修饰 all nodes，分词短语 coordinating and supplying all resources 作定语修饰 specialized nodes。

译文：网络可以由一起平等工作的所有节点来控制，或者是由协调和提供所有资源的专用节点控制。

Exercises

Ⅰ. Put "true" or "false" in the brackets for the following statements according to the passage.

1. (　) The network architecture describes how a computer network is arranged; the arrangement is called topology.

2. (　) In a distributed processing system, computing power is located and shared at the same location.

3. (　) In a star network, each device can require to send message simultaneously.

4. (　) In a bus network, as information passes through the bus, every node can receive it.

5. (　) With microcomputers, the ring arrangement is the most frequently used of the four networks.

6. (　) A hierarchical network is useful in distributed organizations.

7. (　) In a peer-to-peer network system nodes can act as both servers and clients, that is, the nods can exchange their roles.

8. (　) In a client/server system, all nodes on the network have equal responsibilities for coordinating the network's activities.

9. (　) The client/server system has powerful network management software.

10. (　) One organization can only have one network configuration.

Ⅱ. Fill in the blanks according to the passage.

1. Network architecture describes how computer network is _____ and how computer resources are _____.

2. A _____ is any device that is connected to a network.

3. Network operating system _____ and _____ electronic communication, information, and resource sharing.

4. In a star network, control is maintained by _____.

5. In a ring network, each device is connected to two other devices, forming a _____.

6. A ring network is useful in a _____ organization because it makes possible a _____ data processing system.

7. _____ network systems use one powerful computer to coordinate and supply services to all other nodes on the network.

8. The most common network strategies are _____ and client/server systems.

9. Every network has a _____ or way of coordinating the sharing of information and resources.

10. The hierarchical network consists of several computers linked to a _____ _____ host computer, just like a star network.

Ⅲ. Translate the following words and expressions into Chinese.

1. distributed processing system　　　　　　2. network architecture

3. peer-to-peer system
4. strategy
5. hierarchical network
6. host computer
7. peripheral device
8. decentralized organization
9. self-contained
10. configuration

2.4.1 Reading Material

The Growth and Development of the Internet

This section consists of brief entries, showing the year of an event, a comment regarding the event and the personalities, as appropriate.

1969: The first four nodes of the ARPANET are deployed. In order, they come up at UCLA, SRI, university of California at Santa Barbara, and the University of Utah.

1972: Ray Tomlinson of BBN introduces network email and the @ sign.

1972: Norm Abramson's Alohanet connected to the ARPANET. This eventually led to the Packet Radio Net (PRNET) and was the first additional network connected to the ARPANET.

1973: Motivated by the three interconnected networks, Bob Kahn and Vint Cerf conceive of the Transmission Control Protocol (TCP) and publish the idea formally in 1974. This architecture would allow packet networks of different kinds to interconnect and machines to communicate across interconnected networks.

1977: TCP is used to connect three networks (ARPANET, PRNET, and SATNET) in an intercontinental demonstration.

1979: CSNET is conceived as a result of a meeting convened by Larry Landweber. The National Science Foundation (NSF) funds it in early 1981. This enabled the connection of many more computer science researchers to the growing Internet.

1980: Ethernet goes commercial through 3-Com and other vendors.

1981: IBM introduces their first personal computer (PC).

1983: TCP/IP becomes the official standard for the ARPANET.

1984: The Domain Name System (DNS) is designed by Paul Mockapetris.

1988: Robert Morris unleashes the first Internet worm. This is the commencement of the dark side of the Internet.

1989: Tim Berners-Lee proposes a global hypertext project, to be known as the World Wide Web (WWW).

1989: ARPANET backbone replaced by NSFNET.

1991: Tim Berners-Lee makes the first Web site available on the Internet.

1992: Internet Society is formed.

1993: The Mosaic browser is released by Marc Andreessen and Eric Bina of the National Center for Supercomputer Applications (NCSA) at the University of Illinois, Urbana-Champaign.

1994: Netscape browser is released.

1995: Bill Gates issues "The Internet Tidal Wave" memo within Microsoft.

1996: In the United States, more email is sent than postal mail.

1998: Blogs begin to appear.

1998: Voice over IP (VoIP) equipment begins rolling out.

2001: English is no longer the language of the majority of Internet users. It falls to a 45 percent share.

2002: Broadband users exceed the number of dial-up users in the United States.

2005: 812 million cell phones sold; 219 million laptop computers sold.

2005: Google is the darling of the Internet.

2005: Peer-to-peer networks grow.

2005: Google Maps and Google Earth appear.

2005: Web 2.0 technologies heat up.

2007: Mobile TV ads, applications, and content emerging.

2007: Apple introduces the iPhone.

2007: Google lays out Android, its open cell phone platform.

It is clear that the Internet is a vital force and has grown considerably over its lifetime.

Words

unleash	*v.*	释放
commencement	*n.*	开始，毕业典礼
tidal	*adj.*	潮汐的，定时涨落的
postal	*adj.*	邮政的，邮局的
vital	*adj.*	重大的，生机的，至关重要的
lifetime	*n.*	一生，终生，寿命
entry	*n.*	登录，条目，进入，入口
personality	*n.*	个性，人格，名人
appropriate	*adj.*	适当的

Abbreviations

UCLA(University of California at Los Angeles)　　(美国)加利福尼亚大学洛杉矶分校

2.4.2　正文参考译文

网络体系结构

网络体系结构描述计算机网络是如何连接以及计算机资源是如何共享的。

描述计算机网络的专业术语很多，经常用于描述的有：节点、客户机、服务器、网络操作系统、分布处理及主计算机。

节点是连接到网络的任一设备，它可以是计算机、打印机、通信或数据存储设备。

客户机是请求和使用其他节点资源的节点。通常情况下，客户机就是用户的微机。

服务器是和其他节点共享资源的节点。根据所共享的资源不同，服务器可以被称为文件服务器、打印机服务器、通信服务器或数据库服务器。

网络操作系统像 Windows，它控制和协调网络中计算机间的活动。这些活动包括电子通信、信息以及资源共享。

在分布式处理系统中，计算能力被分布在不同的地方共享。这类系统在分散型机构中很常见，其分开的办公室具有它们自己的计算机系统。这些分开的办公室的计算机系统联网到该机构的主计算机或中央计算机。

主计算机是大的中央计算机，通常是小型机或大型机。

网络可以仅由微机组成，也可以由微机或其他设备与较大计算机结合而来。网络可以由一起平等工作的所有节点来控制，或者是由协调和提供所有资源的专用节点控制。网络可以是简单的，也可以是复杂的；可以是独立的，也可以是分散在大的地理区域内的。

结构　网络可以用几种不同的方式排列或连接。主要的结构有四种：星型、总线型、环型和层次型。

在星型网络中，一些小的计算机或外部设备被连接到一个中心设备。这个中心设备可以是主计算机或文件服务器，所有的通信都通过这个中心设备。控制是由轮询实现的。也就是说，每一个连接的设备都被询问是否有信息发送，然后每个设备被轮流允许发送其信息。星型网络最大的优点是它能用于提供分时系统，即几个用户可以在一个中心计算机上共享资源("时间")。在将几个微机连接到大型机上以访问一个机构的数据库时，经常使用星型结构排列。

在总线型网络中，网络中的每一个设备处理自己的通信控制，该网络中没有主计算机。所有通信沿着被称为总线的公共连接电缆传输。信息通过总线时，由每一个设备检验以判定该信息是否是给自己的。当只有几个微机需要连接在一起时，一般使用总线型网络。这种结构常见于发送电子邮件或共享存储在不同微机上的数据的系统。对于共享资源，总线型网络没有星型网络的效率高(这是因为总线型网络不是直接连接到资源上的)。然而，总线型网络成本不高而且很常用。

在环型网络中，每一个设备被连接到其他两个设备形成环路。环型网络中没有中央文件服务器或中央计算机，信息通过环路直至到达正确的目的地。对于微型机来讲，环型结构是这四种网络当中使用最少的。然而环型网经常被用于连接大型机，尤其是在跨度很大的地理区域。这些大型机趋于非常自主的运作，它们完成自己绝大多数或全部的处理任务，偶尔与其他大型机共享数据和程序。环型网络在分散型机构中很有用，因为这使得分布式数据处理系统成为可能，即计算机在它们自己分散的位置就可以完成处理任务。然而，它们也可以相互共享程序、数据以及其他的资源。

层次型网络是由几个连接到中心计算机的计算机构成的，就像星型网络。然而，这些计算机又可以作为其他更小的计算机或外部设备的主机。这样，在层次顶部的主计算机就可能是大型机，在大型机下层的可能是小型机，小型机下层是微机。层次型网络(也被称为混合型网络)允许各种计算机共享数据库、处理能力及不同的输出设备。层次型网络适合于集中的机构。例如，一个机构内部的不同部门可以有连接到部门小型机上的单个微机，这个小型机依次可以连接到公司的大型机，公司大型机包含所有机器可以访问的程序和数据。

策略　每一个网络都有协调信息和资源的策略或方法，最常用的网络策略是对等和客户/服务器系统。

在对等网络系统中，节点既可以是服务器也可以是客户机。例如，一个微机可以获得位于另一个微机上的文件,也可以给其他的微机提供文件。对等系统的典型结构是总线型网络。常用的网络操作系统有 Apple 公司的 Macintosh Peer-to-Peer LANS、Novell 公司的 Netware Lite 和 Microsoft 公司的 Windows for Workgroups。使用这类策略有几个优点。这种网络不贵且易于安装,通常在节点数小于 10 个的小系统中能较好地工作。然而,当节点的数目增加时,网络的性能就会下降。缺点是缺乏强大的管理软件,以有效地监控大的网络活动。由于这些原因,对等网络一般用于小型网络中。

客户/服务器网络系统使用一台功能强大的计算机来协调网络上的其他节点,并提供服务。这一策略是基于专业化的,服务器节点协调和提供专门的服务,客户节点请求服务。常用的网络操作系统有:Novell 公司的 Netware、Microsoft 公司的 LAN 和 Windows NT。客户/服务器网络系统的优点是能够有效地处理大的网络活动,同时它还有强有力的网络管理软件监视和控制网络活动。其缺点是安装和维护的费用较高。

2.4.3 阅读材料参考译文

互联网的成长和发展

本部分是一些简单条目,包括事件的年份、关于事件的注释以及人物。

1969:具有 4 个节点的 ARPANET 装备好了,分别位于加利福尼亚大学洛杉矶分校(UCLA)、SRI、加利福尼亚大学圣迭戈分校(UCSB)和犹他大学。

1972:BBN 的 Ray Tomlinson 引入网络电子邮件和@符号。

1972:Norm Abramson 的 Alohanet 连接到 ARPANET(阿帕网),成为第一个连接到阿帕网的其他网络,这最终促使了分组无线网(PRNET)的形成。

1973:由三个互联网络推动,Bob Kahn 和 Vint Cerf 提出传输控制协议(TCP)并于 1974 年正式公布。这种架构允许不同类型的包交换网络互连,机器能通过互连的网络进行通信。

1977:TCP 被用于连接洲际示范的三个网络(阿帕网 ARPANET、分组无线网 PRNET 和卫星通信网 SATNET)。

1979:CSNET 被认为是在 Larry Landweber 主持召开的一次会议上构思出来的。1981 年初得到美国国家科学基金会(NSF)的资助,使得更多的计算机科学研究人员可以连接到互联网。

1980:以太网通过 3-COM 和其他厂商进入商业市场。

1981:IBM 公司推出了他们的第一台个人计算机(PC)。

1983:TCP/IP 协议成为阿帕网(ARPANET)的正式标准。

1984:Paul Mockapetris 提出域名系统(DNS)。

1988:Robert Morris 释放第一个互联网蠕虫病毒,这是互联网黑暗面的开始。

1989:Tim Berners-Lee 提出了被称为万维网(WWW)的全球超文本项目。

1989:阿帕网(ARPANET)的骨干作用被 NSFNET 取代。

1991:Tim Berners-Lee 建成了互联网上的第一个网站。

1992:互联网协会成立。

1993:国家超级计算机应用中心(NCSA)的 Marc Andreessen 和 Eric Bina 在伊利诺斯大学 Urbana-Champaign 分校发布了 Mosaic 浏览器。

1994:Netscape 浏览器发布。

1995：比尔·盖茨在微软发表了"互联网浪潮"备忘录。
1996：在美国，电子邮件发送量超过邮寄的邮件。
1998：博客开始出现。
1998：网络电话(VoIP)设备开始推出。
2001：英语已经不再是大多数互联网用户的语言，下降到45%的份额。
2002：在美国宽带用户的数量超过拨号上网用户的数量。
2005：手机销量达8.12亿部；笔记本电脑销量达2.19亿。
2005：谷歌成为互联网的宠儿。
2005：点对点网络增长。
2005：Google Maps和Google Earth出现。
2005：Web 2.0技术升温。
2007：移动电视广告、应用和内容出现。
2007：苹果公司推出iPhone。
2007：谷歌投资开放式手机平台Android。

显然，互联网是一个重要力量，从出现以来已经显著成长。

Chapter 3   Internet Applications

3.1 Browsers and E-mails

Want to communicate with a friend across town, in another province, or even in another country? The Internet and the Web are the 21st-Century information resources designed for all of us to use.

Browsers are programs that provide access to Web resources. This software connects you to remote computers, opens and transfers files, displays text and images, and provides in one tool an uncomplicated interface to the Internet and Web documents. Two well-known browsers are Netscape Navigator and Microsoft Internet Explorer. For browsers to connect to other resources, the location or address of the resources must be specified. These addresses are called Uniform Resources Locators (URLs). Following the Domain Name System (DNS), all URLs have at least three basic parts. The first part presents the protocol used to connect to the resource. The protocol http:// is by far the most common. The second part presents the domain name or the name of the server where the resource is located. The server is identified as www.aol.com. (Many URLs have additional parts specifying directory paths, file names, and pointers) The last part of the domain name following the dot (.) is the domain code. It identifies the type of organization. For example, com indicated a commercial site.

The URL http://www.aol.com connects your computer to a computer that provides information about America Online (AOL). These informational locations on the Web are called Web sites. Moving from one Web site to another is called surfing.

Once the browser has been connected to a Web site, a document file is sent to your computer. This document contains Hypertext Markup Language (HTML) commands. The browser interprets the HTML commands and displays the document as a Web page. Typically, the first page of a Web site is referred to as its home page. [1]The home page presents information about the site along with references and hyperlinks, or connections to other documents that contain related information such as text files, graphic images, audio, and video clips.

These documents may be located on a nearby computer system or on one halfway around the world. The references appear as underlined and colored text and /or images on the Web page. To access the referenced material, all you do is click on the highlighted text or image. A link is automatically made to the computer containing the material, and the referenced material appears.

Communication is the most popular Internet activity. The impact of electronic communication cannot be overestimated. At a personal level, friends and family can stay in contact with one another even when separated by thousands of miles. [2]At a business level, electronic communication has become standard and many times preferred way to stay in touch with suppliers, employees, and customers.

You can communicate with anyone in the world who has an Internet address or e-mail account with a system connected to the Internet. All you need is access to the Internet and an e-mail program. Two of the most widely used e-mail programs are Microsoft's Outlook Express and Netscape's Navigator.

A typical e-mail message has three basic elements: header, message and signature. The header appears first and typically includes the following information:

- Addresses: Addresses of the persons sending, receiving, and, optionally, anyone else who is to receive copies.
- Subject: A one-line description, used to present the topic of the message. Subject lines typically are displayed when a person checks his or her mail-box.
- Attachments: Many e-mail programs allow you to attach files such as documents and worksheets. If a message has an attachment, the file name appears on the attachment line.

The letter or message comes next. It is typically short and to the point. Finally, the signature line provides additional information about the sender. Typically, this information includes the sender's name, address, and telephone number.

E-mail addresses have two basic parts. The first part is the user's name and the second part is the domain name, which includes the domain code.

You can also use e-mail to communicate with people you do not know but with whom you wish to share ideas and interests. You can participate in discussions and debates that range from general topics like current events and movies to specialized forums like computer troubleshooting and Star Trek.

Mailing lists allow members of a mailing list to communicate by sending messages to a list address. Each message is then copied and sent via e-mail to every member of the mailing list. To participate in a mailing list, you must first subscribe by sending an e-mail request to the mailing list subscription address. Once you are a member of a list, you can expect to receive e-mail from other on the list. You may find the number of messages to be overwhelming. If you want to cancel a mailing list, send an e-mail request to "unsubscribe" to the subscription address.

Words

attachment	*n.*	附件
debate	*v.*	辩论
document	*n.*	文件，公文，证件
domain	*n.*	域，领域
forum	*n.*	论坛
halfway	*adv.*	在半途
hyperlink	*n.*	超链接

impact	v.	影响
indicate	v.	显示，象征，指示，指出
interpret	v.	解释，翻译，理解
overestimate	v.	过高估计
overwhelm	v.	压倒，使不知所措
signature	v.	签名
specialized	adj.	专用的，专门的
subject	n.	主题
subscription	n.	预约，用户，订阅费
surf	v.	冲浪
well-known	adj.	著名的，有名的

Phrases

at least	至少
along with	和……一起，伴随……
access to	进入，有权使用
by far	到目前为止
connect to	与……相连
in touch	联系，接触
to the point	中肯，扼要

Abbreviations

AOL (America Online)	美国在线
DNS (the Domain Name System)	域名系统
HTML (Hypertext Markup Language)	超文本语言
URL (Uniform Resource Locator)	统一资源定位符

Notes

[1] 例句：The home page presents information about the site along with references and hyperlinks, or connections to other documents that contain related information such as text files, graphic images, audio, and video clips.

分析：本句是复合句，information 和 connections 是并列宾语，that contain …是定语从句，修饰 documents。

译文：主页显示有关站点的信息、参考及超链接，或者与包含相关信息(如文件夹、图像、音频和视频)的其他文件的链接。

[2] 例句：At a business level, electronic communication has become standard and many times preferred way to stay in touch with suppliers, employees, and customers.

分析：本句是简单句，standard 和 preferred way 是 has become 的并列宾语，to stay in touch with suppliers, employees, and customers 是动词不定式作定语，修饰

preferred way。

译文：对于企业用户来说，电子通信已经成为标准的、常采用的方法，以保持与供应商、雇员和用户的联系。

Exercises

Ⅰ. Put "true" or "false" in the brackets for the following statements according to the passage.

1. ( ) Domain code identifies the location of the resources.
2. ( ) URL is the same as domain name.
3. ( ) HTML commands are commands that display Web pages.
4. ( ) A typical e-mail message has three basic elements: addresses, message and signature.
5. ( ) Following the domain name system e-mail addresses have two basic parts, the first is the user's name and the second is the domain name.
6. ( ) The signature line provides additional information about the receiver.
7. ( ) The first page of a Web site is referred to as its home page.
8. ( ) Moving from one Web site to another is called searching.
9. ( ) Mailing lists allow members of a mailing list to communicate by sending messages to a list address.
10. ( ) To participate in a mailing list at the first time, you just directly send messages to a list address.

Ⅱ. Fill in the blanks according to the passage.

1. _____ are programs that provide access to Web resources.
2. URLs have at least two basic parts, which are _____ and _____.
3. Moving from one Web site to another is called _____.
4. Communication is the most popular Internet activity. Two categories are _____ and _____.
5. Two of the most widely used e-mail programs are _____ and _____.
6. _____ are ways to communicate electronically with one or more individuals.
7. The last part of the domain name following the dot (.) is the _____.
8. Communication is the most _____ Internet activity.
9. To _____ in a mailing list, you must first subscribe by sending an e-mail request to the mailing list subscription address.
10. If you want to _____ a mailing list, send an e-mail request to "unsubscribe" to the subscription address.

Ⅲ. Translate the following words and expressions into Chinese.

1. browser   6. DNS
2. URL   7. hyperlinks
3. protocol   8. site
4. signature   9. connect to
5. HTML   10. Web page

Chapter 3　Internet Applications

3.1.1　Reading Material

Online Learning Grows, but Research Finds Mixed Results

Forty-two of the fifty American states offered some kind of public online learning this past school year. One state, Michigan, now requires all students to have an online learning experience before they finish high school.

Even the idea of a school has changed since the rise of the Internet in the nineteen nineties.

A new report from the Center for Evaluation and Education Policy at Indiana University says eighteen states have full-time virtual schools. There are no buildings. All classes are online.

Online learners might work at different times. But there might be set times for class discussions—by text, voice or video—and virtual office hours for teachers.

Florida started the first statewide public virtual school in the United States in nineteen ninety-seven.

Today, the Florida Virtual School offers more than ninety courses. Fifty-six thousand students were enrolled as of December. Almost sixty percent were female. The school's Web site says each student was enrolled in an average of two classes.

Two-thirds were also enrolled in public or charter schools. Charter schools are privately operated with public money. Other students are home-schooled or in private school.

Florida Virtual School has now opened the Florida Virtual Global School. Students in other countries pay for classes. Janet Heiking teaches an English class. She lives in Indianapolis, Indiana. Her students live as far away as Africa and Japan.

She says they are taking her Advanced Placement class to prepare for attending an American college. They can earn college credits by passing the A.P. test.

So how good are virtual schools? Studies have shown mixed results, as that new report from Indiana University notes.

For example, students at Florida Virtual School earned higher grades than those taking the same courses the traditional way. And they scored higher on a statewide test.

But virtual school students in Kansas and Colorado had lower test scores or performed at a lower level than traditional learners.

Studies also find that virtual schools may not save much in operating costs.

Education experts say the mixed results suggest the need for more research to find the best ways to teach in virtual schools. Also, they say schools of education need to train more teachers to work in both physical and virtual classrooms.

Words

charter	*n.*	(由统治者或政府发给城镇或大学等的)特许状，契约，执照
Colorado	*n.*	美国科罗拉多州(美国州名)
credit	*n.*	信任，信用，[财务]贷方，银行存款，学分

enroll	v.	登记，招收，使入伍(或入会、入学等)，参加
Florida	n.	佛罗里达(美国州名)
Indiana	n.	印地安那州(美国州名)
Kansas	n.	堪萨斯州(美国州名)
Michigan	n.	密歇根州(美国州名)
mixed	adj.	混合的
online	adj.	[计]联机的
statewide	adj.	遍及全州的，全州范围的

Phrases

Advanced Placement   跳级(生)，美国在高中阶段开设的具有大学水平的课程

3.1.2 正文参考译文

浏览器和电子邮件

想要跨城市、省甚至国家与一位朋友进行信息交流么？Internet 和 Web 是为此设计的 21 世纪信息资源。

浏览器是提供访问 Web 资源的程序。该软件把我们和远程计算机连接起来，利用它可以打开以及传送文件，显示正文和图像，还可以给 Internet 和 Web 资源提供一个简单接口。Netscape 公司的 Navigator 和微软公司的 IE 是两个有名的浏览器。要链接其他资源的浏览器，必须先指定资源的位置或地址，这些地址被称为统一资源定位符(URL)。继域名系统(DNS)之后，所有的 URL 至少有三个基本部分：第一部分表示链接资源的协议，协议 http:// 是当今最普遍的；第二部分表示资源所在的域名或服务器名，一个服务器可能被标识为 www.aol.com(很多 URL 还有附加部分，用于指定目录路径、文件名和指针)；在域名中圆点之后的最后一个部分是域码(顶级域名)，用于标明机构类型，例如，.com 表示该网站是商业网站。

"http://www.aol.com"把用户计算机和提供关于"美国在线(AOL)"信息的计算机连接起来。Web 上这些信息的位置被称为网站，而从一个网站移到另一个网站称为"冲浪"。

一旦我们连接上一个网站，就会有文档发送到我们所用的计算机。这个文档包含 HTML 命令。浏览器解释 HTML 命令并以网页的形式显示。通常，一个网站的第一页被称为主页。主页显示有关站点的信息、参考及超链接，或者与包含相关信息(如文件夹、图像、音频和视频)的其他文件的链接。

这些文档可能位于地理位置离我们不远的计算机系统中，也可能在世界各地的任何一个角落。在 Web 页中，参考以彩色文本或图片加下画线的方式出现。要访问参考资料，单击醒目的文本和图片，即可以自动链接到包含该信息的计算机，显示出详细的参考资料。

通信是 Internet 最常用的活动之一。电子通信的范围越来越大。对于个人用户来说，朋友和家人能够远隔千里而保持联络。对于企业用户来说，电子通信已经成为标准的、常采用的方法，以保持与供应商、雇员和用户的联系。

访问 Internet 和邮件程序，就能够和世界上任何一个拥有 Internet 地址或者 E-mail 账户的联网的人交流。两个最常用的邮件程序是微软的 Outlook Express 和 Netscape 的 Navigator。

E-mail 信息有三个主要构成部分：抬头、信息和签名。抬头出现在 E-mail 最前面，主要包含以下的信息。

- 地址：收信和发信人的地址，也可以加上抄送人的地址。
- 主题：表示邮件主题的一行描述信息。当一个人查看邮箱时，会先看到主题内容。
- 附件：许多邮件程序允许附加文档或工作表。如果邮件有附件，其文件名会在附件行中显示。

接下来就是信件或信息，一般比较简短并且切入主题，最后签名部分提供一些发送者的其他信息，主要包含发送者的姓名、地址和电话号码。

E-mail 地址由两个基本部分组成。第一部分是用户名，第二部分是域名，包括域码(顶级域名)。

使用 E-mail，可以和一个不认识但愿意与之交流思想和兴趣爱好的人通信；也可以参加别人的讨论，话题从一般的当前时事或电影，到专门的计算机检修和 Star Trek 之类。

邮件列表允许列表用户通过向一组列表地址发送信息来交流。每条信息可以复制并通过 E-mail 发往列表中的每个成员。想加入邮件列表，必须先向预定地址通过 E-mail 提交申请预定。一旦成为列表成员，就会收到其他列表成员发送的 E-mail，这样就可能会发现信息多得令人不知所措。如果想取消邮件列表，则必须向预定地址发送 E-mail 取消预定。

3.1.3 阅读材料参考译文

网络学习在发展，但研究发现结果不同

在过去的一学年，美国 50 个州中有 42 个州提供某些公众网络学习课程，密西根州现在要求所有学生在高中阶段要有网络学习的经验。

随着 20 世纪 90 年代 Internet 的发展，学校的概念也在变化。

印第安那州大学评估与教育政策中心给出的最新报告指出，18 个州有全日制虚拟学校，这些学校没有校舍，所有课程都是网络课程。

网络学习者可以在不同的时间段学习，但是会在特定的时间通过文字、声音或视频进行课堂讨论，也有教师的虚拟工作时间。

佛罗里达州在 1997 年建立了美国第一个全州范围的公立虚拟学校。

现在，佛罗里达虚拟学校提供 90 多门课程。56 000 学生 10 月份注册入学。约 60%的学生是女生。学校网站统计显示平均每个学生注册学习两门课程。

三分之二的学生也参加公立或特许学校。特许学校是用公共资金私人经营的，其他学生在家中或私立学校接受教育。

佛罗里达虚拟学校现已开放成为佛罗里达虚拟全球学校。其他国家的学生为课程付费。Janet Heiking 任教英语课程。她住在印第安纳州印第安纳波利斯，她的学生有的远在非洲和日本。

她说，他们正在学习 AP 课程，准备进入美国的学院。通过 AP 考试他们就可得到学院的学分。

那么网络学校怎么样呢？印第安那州立大学的最新研究报告给出了不同的结果。

佛罗里达虚拟学校的学生取得的成绩高于校内传统学习方式的学生在同一门课程取得的成绩。并且在全州范围的考试中，虚拟学校的学生取得的成绩也更好。

但是堪萨斯州和科罗拉多州的虚拟学校的学生成绩就低于传统学习方式的学生成绩。研究还发现虚拟学校并不能节约许多运行成本。

教育专家指出这种不同的结果提示应该进行更多的研究工作，以找出虚拟学校的最好教学方法，并且教育学院应该培训更多在真实和虚拟的教室都可进行教学的教师。

3.2 Search Tools

The Web can be an incredible resource providing information on nearly any topic imaginable. Are you planning a trip? Writing an economics paper? Looking for a movie review? Trying to locate a long-lost friend? Information sources related to these questions, and much, much more are available on the Web.

With over two billion pages and more being added daily, the Web is a massive collection of interrelated pages. With so much available information, locating the precise information you need can be difficult. Fortunately, a number of organizations called search services or search providers can help you locate the information you need. They maintain huge databases relating to information provided on the Web and the Internet. The information stored at these databases includes addresses, content descriptions or classifications, and keywords appearing on Web pages and other Internet informational resources. Special programs called agents, spiders, or bots continually look for new information and update the search services databases. Additionally, search services provide special programs called search engines that you can use to locate specific information of the Web.

Search Engines

[1]Search engines are specialized programs that assist you in locating information on the Web and the Internet. To find information, you go to the search service's Web site and use their search engine. Yahoo's search engine, like most others, provides two different search approaches.

- Keyword Search

In a keyword search, you enter a keyword or phrase reflecting the information you want. The search engine compares your entry against its database and returns a list of hits or sites that contain the keywords. Each hit includes a hyperlink to the referenced Web page (or other resource) along with a brief discussion of the information contained at that location. Many searches result in a large number of hits. For example, if you were to enter the keyword travel, you would get over a thousand hits. [2]Search engines order the hits according to those sites that most likely contain the information requested and present the list to you in that order, usually in groups of ten.

- Directory Search

Most search engines also provide a directory or list of categories or topics such as Arts &Humanities, Business & Economics, Computers & Internet. In a directory search, also known as index search. You select a category that fits the information that you want. Another list of subtopics relates to the topic you selected appears. You select the subtopic that best relates to your topic and another subtopic list appears. You continue to narrow your search in this manner until a list of Web sites appears. This list corresponds to the hit list previously discussed.

As a general rule, if you are searching for general information, use the directory search approach. For example, to find general information about music, use a directory search beginning with the category Arts &Humanities. If you are searching for specific information, use the key word approach. For example, if you were looking for a specific MP3 file, use a keyword search entering the album title and/or the artist's name in the text selection box.

A recent study by the NEC Research Institute found that any one search engine includes only a fraction of the informational sources on the Web. Therefore, it is highly recommended that you use more than one search engine when researching important topics. Or, you could use a special type of search engine called a metasearch engine.

Metasearch Engines

One way to research a topic is to visit the Web site for several individual search engines. At each site, enter the search instructions, wait for the hits to appear, review the list, and visit selected sites. This process can be quite time-consuming and duplicate responses from different search engines are inevitable. Metasearch engines offer an alternative.

Metasearch engines are programs that automatically submit your search request to several search engines simultaneously. The metasearch engine receives the results, eliminates duplicates, orders the hits, and then provides the edited list to you. There are several metasearch sites available on the Web. One of the best known is Metacrawler.

Specialized Search Engines

Specialized search engines focus on subject-specific Web sites. Specialized sites can potentially save you time by narrowing your search. For example, let's say you are researching a paper about the fashion industry. You could begin with a general search engine like Yahoo! Or, you could go to a search engine that specialized specifically in fashion.

Words

agent	n.	代理人，情报人
album	n.	专辑唱片簿，相片簿
brief	adj.	简短的，短暂的
correspond	v.	符合，一致
database	n.	数据库
duplicate	n.	完全相同的副本
entry	n.	进入，登记，条目
fraction	n.	小部分，片断，分数
hit	n.	命中(指两个数据项的成功比较或匹配)

	v.	打，击中
imaginable	adj.	可能的，可想象的
incredible	adj.	难以置信的
keyword	n.	关键字，关键词
massive	adj.	粗大的，巨大的
precise	adj.	精确的，准确的
simultaneously	adv.	同时地
spider	n.	蜘蛛
specific	adj.	清楚的，明确的
update	v.	升级

Phrases

as a rule	通常，一般(来说)
assist in	帮助……参加……
compare against	和……相比较
fashion industry	时装工业
get over	克服(困难等)
look for	搜索
relate to	与……有关，涉及
result in	引起，导致，产生……的结果

Abbreviations

MP3	一种音频压缩格式
NEC (Nippon Electric Company)	日本电器公司

Notes

[1] 例句：Search engines are specialized programs that assist you in locating information on the Web and the Internet.

分析：本句为复合句，that assist you in locating information on the Web and the Internet 是定语从句，修饰 specialized programs。英译汉时，定语从句不一定翻译成定语，可采用分译法。

译文：搜索引擎是专用的程序，能帮助用户定位 Web 和 Internet 上的信息。

[2] 例句：Search engines order the hits according to those sites that most likely contain the information requested and present the list to you in that order, usually in groups of ten.

分析：本句是复杂句，order 和 present 是主句中并列的谓语动词。

译文：搜索引擎根据需求信息与网站内容的匹配程度对搜索结果进行排序，并且按照这个顺序将相关信息列出来，通常每组 10 个。

Chapter 3 Internet Applications

Exercises

Ⅰ. Put "true" or "false" in the brackets for the following statements according to the passage.

1. ( ) Usually, many searches result in a large number of hits, and search engines list them randomly on the Web site.
2. ( ) Bots are also known as agents and spiders.
3. ( ) Most search engines provide only one search approach, namely keyword search.
4. ( ) Keyword search is more convenient than directory search.
5. ( ) Search engines help you locate information on the Web.
6. ( ) Specialized search engines focus on subject-specific Web sites.
7. ( ) Each hit includes a hyperlink to the referenced Web page along with a brief discussion of the information contained at that location.
8. ( ) In order to use keyword search, you only need to enter word like "music", and search engine will list all interrelated information from databases.
9. ( ) In a keyword search, a keyword is entered and a list of hits or sites containing the keywords is presented, and usually those duplicate sites are already eliminated.
10. ( ) Say, you are researching a paper about the cooking, you'd better begin with a search engine that specializes in that, for it may save time than with a general search engine like Yahoo!.

Ⅱ. Fill in the blanks according to the passage.

1. The _____ can be an incredible resource providing information on nearly any topic imaginable.
2. Bots are used for looking for new information and _____ the search services' databases.
3. _____ are specialized programs that assist you in locating information on the Web and the Internet.
4. In a directory or _____ search, you select a category that fits the information you want.
5. As a general rule, keyword search is good for _____ information, while directory search is good for _____ information.
6. _____ engines are programs that automatically submit search requests to several search engines simultaneously.
7. _____ are the list of sites that contain the keywords of a keyword search.
8. Metasearch engine receives the results, _____ duplicate sites, _____ hits, and then provides the edited list to you.
9. Search services maintain _____ and provide search engines to _____ information.
10. Search services maintain huge _____ relating to information provided on the Web and the Internet.

Ⅲ. Translate the following words and expressions into Chinese.

1. search engine   2. keyword search

3. directory search
4. duplicate
5. specialized search engine
6. spider
7. NEC
8. fashion industry
9. update
10. bot

3.2.1 Reading Material

The Google

Google is now the most dominant search tool on the web, setting the standards that others try to follow and better, as yet unsuccessfully. It was founded in 1998—relatively late compared to many of the popular search engines—by Larry Page and Sergey Brin, who were graduate students from Stanford University.

Page and Brin had been working together on a search engine they called "BackRub" since early 1996, but with the encouragement of Yahoo! co-founder David Filo, they decided to start a company in 1998 and went looking for investors to back them. Google, Inc. was established on September 7, 1998. The founders hired Craig Silverstein—who was later to become Director of Technology—as their first employee, and started the business in a friend's garage. Google is derived from the word 'googol' which stands for the number 1 followed by 100 zeros. Sergey Brin and Larry Page chose this as they thought that was appropriate for a search engine that would seek out information from the astronomical number of possible sources contained on the Internet.

Google was still in an alpha stage, with an index of just 25 million pages, but it was handling 10,000 search queries every day. The search engine and the company grew quickly through word of mouth, initially with regular web users coming across the tool and finding the results to their liking.

Usage spread rapidly through press coverage, awards and recommendations, whilst Google's effectiveness and relevance, its speed and reliability, plus clean visual effects and 'quirky' nature all contributed to a rapid increase in the number of new advocates.

Google took a major step forward in 2000 when it replaced Inktomi as the provider of supplementary search results on Yahoo. Following this it won further successes and provided search data to Yahoo as its primary results, as well as to AOL, Netscape, Freeserve and BBCi in the UK. This gave Google exceptional coverage of web searches and established its reputation as one of the most reliable and accurate search tools, making it the clear market leader.

Despite losing the Yahoo relationship in 2004, Google continued to increase its coverage of the web search market and developed numerous regional versions of its search tool, both in English and other languages, so that its global dominance grew.

Google has also been actively developing a range of search options, including an image search, news search, shopping search (Froogle) and local search options. In addition, following Google's IPO in early 2005 it has set itself on a course for Internet domination and to challenge the position of Microsoft as the leading provider of computer services. There has been a series of announcements of new products, including the email service Gmail, the impressive Google Earth product, Google Talk to compete in the growing VoIP market, Google Base and Google Book Search, which is part of its ambitious project to make the content of thousands of books searchable online.

Google has become synonymous with search and has entered the dictionary as a verb—'google something'. The expansion and integration of all Google's different services is making it a dominant player in the online market, but to many websites, Google is also the ultimate ranking target that will make a significant difference between the volumes of traffic received from prospective customer searching the Web.

Words

advocate	n.	提倡者，鼓吹者
	v.	提倡，鼓吹
alpha	n.	希腊字母的第一个字母
ambitious	adj.	有雄心的，野心勃勃的
astronomical	adj.	天文学的，庞大无法估计的
award	n.	奖，奖品
back	v.	后退，支持
course	n.	过程，经过，进程，课程
coverage	n.	覆盖，保险范围，新闻报道(范围)，报道
derive	v.	来自，起源
dominant	adj.	有统治权的，占优势的，支配的
effectiveness	n.	效力，有效，效率，效果；有效性
employee	n.	职工，雇员，店员
exceptional	adj.	例外的，异常的
expansion	n.	扩充，开展，膨胀，辽阔，浩瀚
found	v.	建立，创办，使有根据
founder	n.	创始人，奠基人
handling	n.	处理
hire	n.	租金，工钱，租用，雇用
	v.	雇请，出租，受雇
initially	adv.	最初，开头
integration	n.	综合
investor	n.	投资者
Netscape		美国 Netscape 公司
press	n.	压，按，印刷，压力，新闻
	v.	压，压榨，逼迫，受压
prospective	adj.	预期的
quirky	adj.	诡诈的，离奇的

relevance	n.	中肯，适当
reputation	n.	名誉，名声
supplementary	adj.	额外的，补充的，(数)补角的
synonymous	adj.	同义的
whilst	conj.	时时，同时

Phrases

come across	偶遇，碰到
free serve	免费服务
word of mouth	口头的，口头传达的
	口头传播

Abbreviations

BBC(British Broadcasting)	英国广播公司
IPO (Initial Public Offering)	初次公开发行(公司股票首度在股市买卖)

3.2.2　正文参考译文

搜索工具

网络拥有令人难以置信的丰富资源，它可以为任何能想得到的主题提供相关信息。你正要旅行吗？在写一篇经济论文吗？或是查找影讯？寻觅多年失去联系的老朋友？与这些问题相关的信息，或是更多的其他信息都可由网络提供。

已有 20 多亿网页并且每天仍在增加，Web 已经成为了一个汇集众多相关网页的巨大宝库。那么，有了这么多信息，想要找到我们所需要的精确信息就显得非常困难了。幸运的是，一些被称为搜索服务或搜索提供商的组织可以帮我们定位所需的信息。它们拥有一个庞大的数据库，所有存储在数据库中的数据都是 Web 或互联网提供的信息，包括地址、内容概述或分类、以及出现在 Web 页面和其他网络资源中的关键字。一些专用的程序，如代理程序、网络蜘蛛程序或网络机器人，它们能够不断地寻找新信息，并升级搜索服务的数据库。另外，搜索服务提供的专用程序(即搜索引擎)，能帮我们定位网页中的具体信息。

搜索引擎

搜索引擎是专用的程序，能帮助用户定位 Web 和 Internet 上的信息。我们可以进入提供搜索服务的 Web 站点使用其搜索引擎来查找信息，例如 Yahoo 的搜索引擎，和其他大多数网站一样，它提供两种不同的搜索方式。

● 关键词搜索引擎

在关键词搜索引擎当中，输入可以反映所查信息的一个关键字或关键词，搜索引擎通过和数据库中的信息进行比较，返回与其匹配的热点链接或者站点。每个热点包含一个带有简要介绍的链接 Web 页或其他资源。许多搜索都会返回大量的热点链接。例如，如果输入关键字"旅游"，就会得到一千多个热点。搜索引擎根据需求信息与网站内容的匹配程度对搜索结果进行排序，并且按照这个顺序将相关信息列出来，通常每组 10 个。

- 目录式搜索引擎

大多数搜索引擎也提供诸如人文艺术、商业经济和计算机 Internet 的目录的分类主题列表。在目录式搜索(也称索引式搜索)引擎中，选择一个与你想要的信息相匹配的类别，与之相关的子类别列表就会显示出来。用户可以继续以这种方式缩小搜索范围，直到列出一系列站点，这个列表与前面讨论过的热点列表相对应。

一般来说，如果只是搜索一般的信息，就可使用目录式搜索引擎。例如，查询关于音乐的信息，以目录式搜索引擎的人文艺术类开始。如果搜索专用的信息，使用关键词搜索引擎。例如，查询一个专用的 MP3 文件，在文本选择框中输入专辑名称和/或艺术家姓名进行关键词搜索。

最近一项 NEC 研究机构的调查表明，任何一个搜索引擎都仅包含 Web 信息资源的一部分。因此，在检索一个重要的主题时，建议你使用多个搜索引擎，或者使用一个特殊的被称为 Metasearch 的搜索引擎。

Metasearch 搜索引擎

要搜寻某个主题还可以是访问多个独立的搜索引擎 Web 站点。在每个站点输入搜索指令，等待反馈的热点链接信息，这样会占用很多时间，且不同的搜索引擎也会返回重复的信息。Metasearch 搜索引擎提供了另外一种方案。

Metasearch 搜索引擎是一个同时自动提交搜索信息给多个搜索引擎的程序，可以接收搜索结果，去除重复部分并且将热点链接排序，最后将编辑好的结果提交给用户。在 Web 上有多个 Metasearch 站点，其中最著名的一个是 Metacrawler。

专用搜索引擎

专用搜索引擎重点查询专用主题的 Web 站点，通过缩小搜索范围来自动节省用户时间。例如，你正研究一篇关于时装工业的论文，就可以从像雅虎(Yahoo)一样的普通搜索引擎开始。或者也可以进入到关于时装的专用搜索引擎。

3.2.3 阅读材料参考译文

谷歌

谷歌是当今最主要的网络搜索工具，其设置的标准至今仍无人超越。谷歌成立于 1998 年——比许多受欢迎的搜索引擎相对较晚，由斯坦福大学的研究生 Larry Page 和 Sergey Brin 建立。

自 1996 年初，Page 和 Brin 开始合作研究搜索引擎，他们称之为"BackRub"。在 Yahoo 联合创始人大卫·费罗的鼓励下，1998 年他们决定建立一个公司，开始寻找支持他们的投资者。谷歌公司成立于 1998 年 9 月 7 日，创始人聘请克雷格弗斯坦——后来成为技术总监——作为其第一个雇员，在一个朋友的车库里开始了经营。谷歌一词来自"googol"，googol 表示 10 的 100 次方。Sergey Brin 和 Larry Page 选择这个词，是因为他们认为这个词适用于一个能够在互联网上天文数目的可能来源中寻找信息的搜索引擎。

谷歌仍在初始阶段，只有 25 000 000 个网页的检索，每天处理 10 000 次搜索查询。搜索引擎和公司通过口头相传迅速成长，最初是网络用户经常碰到的工具，后来成为他们的喜好。

通过新闻报道、奖励和建议，谷歌的使用率迅速增加。而谷歌的有效和实用、快速和可靠性以及清晰的视觉效果和"离奇"的特性都使其新拥护者迅速增多。

2000年，谷歌取代Inktomi作为补充雅虎搜索结果的供应商，又向前迈进了一大步。在此之后取得进一步的成功，为美国在线、网景、Freeserve和英国的BBCi提供数据，也为雅虎提供作为初步结果的搜索数据。这时谷歌网络搜索范围特别大，建立了作为最可靠和准确的搜索工具之一的声誉，成为明确的市场领导者。

2004年，失去了雅虎关系。但是谷歌继续扩大其网络搜索市场的覆盖范围，并且使用英语和其他语言开发搜索工具的地域性版本。因此，谷歌在全球的统治地位不断加强。

谷歌还积极开发了一系列的搜索选项，包括图片搜索、新闻搜索、购物搜索(Froogle)和本地搜索。2005年初首次公开发行股票后，谷歌为自己确定了在互联网的统治地位，并且挑战微软作为计算机服务领先供应商的地位。目前谷歌已经推出了一系列新产品，包括电子邮件服务Gmail、令人印象深刻的谷歌地球产品、在不断扩大的VoIP市场中竞争的谷歌Talk、谷歌基地和谷歌图书搜索。其中图书搜索是其雄心勃勃的部分项目，这样成千上万的图书内容就可以在线搜索。

谷歌已经成为搜索的同义词，并作为一个动词进入字典——google something。所有谷歌不同服务的扩展和集成，使它成为网络市场的大玩家。但是对于许多网站，谷歌也是最终的高级目标，这一目标将使来自潜在客户的网络搜索流量之间出现显著差异。

3.3　Definitions and Content of the Electronic Commerce

Electronic Commerce over the Internet is a new concept. In recent years, it has become so broadly used that it is often left undifferentiated from other current trends which rely on automation, such as concurrent engineering and just in time manufacturing. Many companies, including CyberCash, Dig Cash，First Virtual, and Open Market has provided a variety of electronic commerce services.

[1] If you have access to a personal computer (PC) and can connect to the Internet with a browser, you can do business online. No more worries about programming. No more searching for outdated catalogs as a customer or printing catalogs as a merchant. No more looking for phone numbers, paying long-distance to connect, or keeping the store open late into the evening. Just get on the Web, open an online store, and watch your business grow.

The wired world of business, developed technology, human talent, and a new way of doing business make up today's growing worldwide economy. The backbone of this electronic commerce is the Internet. The wired world is not about technology, it is about information, decision making，and communication. The wired world is changing life for everyone, from the single household to the largest corporation. [2]No business can afford to ignore the potential of a connected economy.

Electronic commerce is an emerging concept that describes the process of buying and selling or exchanging of products, services, and information via computer networks including the Internet. Kalakota and Whinston (1997) define EC from these perspectives:

From a communications perspective, EC is the delivery of information, products/services, or payments over telephone lines, computer networks, or any other electronic means.

From a business process perspective, EC is the application of technology toward the automation of business transactions and work flow.

From a service perspective, EC is a tool that addresses the desire of firms, consumers, and management to cut service costs while improving the quality of goods and increasing the speed of service delivery.

From an on-line perspective, EC provides the capability of buying and selling products and information on the Internet and other on-line services.

The term commerce is viewed by some as transactions conducted between business partners. Therefore, the term electronic commerce seems to be fairly narrow to some people. Thus, many use the term e-business. It refers to a broader definition of EC, not just buying and selling but also servicing customers and collaborating with business partners, and conducting electronic transactions within an organization. According to Lou Gerstner, IBM's CEO: "E-business is all about cycle time, speed, globalization, enhanced productivity, reaching new customers and sharing knowledge across institutions for competitive advantage."

Just like any other type of commerce, electronic commerce involves two parties: businesses and consumers. There are three basic types of electronic commerce.

Business-to-Consumer (B2C): These are retailing transactions with individual shoppers. The typical shopper at Amazon.com is a consumer, or a customer. Oftentimes, this arrangement eliminates the middleman by providing manufacturers direct sales to customers. Other times, retail stores create a presence on the Web as another way to reach customers.

Consumer-to-Consumer (C2C): This category involves individuals selling to individuals. This often takes the form of an electronic version of the classified ads or an auction. Goods are described and interested buyers contact sellers to negotiate prices. [3]Unlike traditional sales via classified ads and auctions, buyers and sellers typically never meet face-to-face. Examples are individuals selling in classified ads and selling residential property, cars, and so on. Advertising personal services on the Internet and selling knowledge and expertise is another example of C2C. Several auction sites allow individuals to put items up for auctions. Finally , many individuals are using internal networks to advertise items for sale or service.

Business-to-Business (B2B): This category involves the sale of a product or service from one business to another. This is typically a manufacturer-supplier relationship. For example, a furniture manufacturer requires raw materials such as wood, paint, and varnish. In B2B electronic commerce, manufacturers electronically place orders with suppliers and many times payment is made electronically.

Many people think EC is just having a Web site, but EC is much more than that. There are dozens of applications of EC such as home banking, shopping in on-line stores and malls, buying stocks, finding a job, conducting an auction, and collaborating electronically on research and development projects. To execute these applications, it is necessary to have supporting information and organizational infrastructure and systems. Figure 3.1 shows that the EC applications are supported by infrastructures, and their implementation is dependent on four major areas(shown as supporting pillars): people, public policy, technical standards and protocols, and other organizations. The EC management coordinates the applications, infrastructures, and pillars.

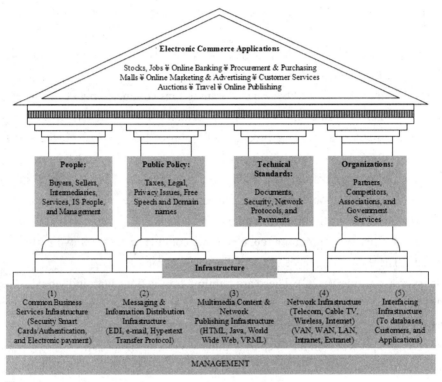

Figure 3.1　Framework for EC

Words

arrangement	n.	安排，准备工作，整理
auction	n.	拍卖
automation	n.	自动化
backbone	n.	脊梁，骨干
broadly	adv.	大体来说
catalogs	n.	目录，一系列
	v.	把……编入目录
collaborate	v.	合作，协作
concurrent	adj.	同时的，和谐的
corporation	n.	公司
coordinate	v.	调整，使协调
define	v.	给(某物)下定义
enhance	v.	增加，提高

execute	v.	执行
household	n.	家庭
	adj.	家用的，普通的
ignore	v.	忽视，不顾，不理
implementation	n.	贯彻，执行
negotiate	v.	议定，商定
outdate	n.	过时的
perspective	n.	远景，观点
pillar	n.	积极支持者，栋梁
potential	adj.	潜在的，有可能的
share	v.	分享，共有，分担
talent	n.	天才，天赋
varnish	n.	清漆
version	n.	描述，说法，版本
virtual	adj.	虚的，实质的

Phrases

afford to	提供，担负得起
have access to	可以到达(使用)，有进入(使用)的机会
look for	寻找
make up	组成，构成，捏造，补充，化妆
rely on	依靠

Abbreviations

B2C(Business To Customer)	企业对顾客(的贸易)
B2B(Business To Business)	企业对企业(的贸易)
CEO(Chief Executive Officer)	首席执行官
EC(Electronic Commerce)	电子贸易

Notes

[1] 例句：If you have access to a personal computer (PC) and can connect to the Internet with a browser, you can do business online.

分析：本句是复合句，have access to...和 can connect to...是条件从句中的并列谓语。

译文：如果你有机会使用一台个人计算机并且能连接上 Internet，你就能在线做生意。

[2] 例句：No business can afford to ignore the potential of a connected economy.

分析：本句是双重否定句。

译文：没有企业能忽视互联网经济的潜能。

[3] 例句：Unlike traditional sales via classified ads and auctions, buyers and sellers typically never meet face-to-face.

分析：本句是简单句。via 是介词，意为"经由"、"通过"，例句：They are going from Shanghai to Los Angeles via Tokyo. 他们从上海经东京去洛杉矶。The news reached us via a friend of mine. 这个消息是通过我的一个朋友传到我们这里的。

译文：与通过分类广告和拍卖方式的传统销售不同，买方与卖方通常从未见过面。

Exercises

Ⅰ. Put "true" or "false" in the brackets for the following statements according to the passage.

1. ( ) You can do business online if you have access to Internet.
2. ( ) The backbone of this electronic commerce is the PC.
3. ( ) Just like any other type of commerce, electronic commerce involves two parties: business and consumer.
4. ( ) B2C involves individuals selling to individuals.
5. ( ) B2B involves the sale of a product or service from one business to another.
6. ( ) From a communications perspective, EC is the application of technology toward the automation of business transactions and workflow.
7. ( ) From a service perspective, EC is a tool that addresses the desire of firms, consumers, and management.
8. ( ) EC cannot provide the capability of buying and selling information.
9. ( ) EC is a broader definition than EB.
10. ( ) EC is just having a Web site.

Ⅱ. Fill in the blanks according to the passage.

1. If you have access to a _____ and can connect to the _____ with a browser, you can do business online.
2. The _____ world of business, where technology, human talent, and a new way of doing business make up today's growing worldwide economy.
3. Electronic Commerce involves two parties, which are _____.
4. B2C involves individuals selling to _____.
5. From a _____ perspective, EC is the _____ of information, products/services, or payments over telephone lines, computer networks, or any other electronic means.
6. From a _____ perspective, EC is the _____ of technology toward the automation of business transactions and work flow.
7. The term commerce is viewed by some as _____ conducted between business partners.
8. _____ refers to a broader definition of EC.
9. According to Lou Gerstner, IBM's CEO: "E-business is all about _____, _____,

_____ enhanced productivity, teaching new customers and sharing knowledge across institutions for competitive advantage."

10. Many people think EC is just having a _____, but EC is much more than that.

III. Translate the following words and expressions into Chinese.

1. the wired world of business
2. B2C
3. manufacturer-supplier
4. automation of business transactions
5. buying and selling products
6. on-line services
7. CEO
8. competitive advantage
9. Web Site
10. home banking

3.3.1 Reading Material

E-Business and Its Advantages

Whether on or off line, customers in today's marketplace want quality products and information in a quick and easy manner. The internet's main benefit is that of speed and convenience. Therefore e-business, which uses the internet as the core for business dealings, can help make a company more customer-friendly in addition to many other things, such as creating a more efficient exchange of information and/or products and services.

1. Removes Location and Availability Restrictions

Users need not be in the same physical location as an e-business and the exchange of information and transactions may take place at any given time, twenty-four hours a day, seven days a week and from any location in the world with Internet access. A physical location is restricted by size and limited to only those customers that can get there, while an online store has a global marketplace with customers and information seekers already waiting in line.

2. Reduces Time and Money Spent

In e-business, there is often a reduction in costs required to complete traditional business procedures. Many of those same traditional business approaches can be eliminated and replaced with electronic means, which are often easier to carry out as well as easier on the pocketbook. For example, compare the cost of sending out 100 direct mailings (paper, postage, staff and all), to sending out bulk e-mail. Also think about the cost of paying rent at a physical location opposed to the cost of maintaining an online site.

3. Heightens Customer Service

With e-business customers receive highly customizable service, and communication is often more effective. There is far more flexibility, availability and faster response times with online support. For example, think about the speed of e-mail inquiries and live chat as opposed to getting on the phone, especially when that business is closed for the day. There is also a faster delivery cycle with online sales, helping strengthen the customer/business relationship. The internet is a powerful channel for reaching new markets and communicating information to customers and partners. Having a better understanding of your customers will help to improve customer satisfaction.

4. Gives Competitive Advantage

The Internet opens up a brand new marketplace to businesses moving online. Competition via the Internet is growing as the Internet itself grows and waiting too long to move online may cause you to lose your place in line entirely. Easy access to real time information is a primary benefit of the Internet, enabling a company to give more efficient and valid information and helping to gain the competitive advantage over those that are not online.

Although there are risks associating with e-business, as with most business decisions, there is also the risk associated with the inability to adapt to the changing times. Change is inevitable in today's marketplace and should be embraced with open arms and open sites!

Words

associate	v.	使联合，交往，结交
availability	n.	可用性，有效性，实用性
benefit	n.	利益，好处
	v.	有益，受益
bulk	n.	大小，体积，大批，大多数，散装
competitive	adj.	竞争的，有竞争力的(价格)
convenience	n.	便利，方便，有益，有用
dealing	n.	行为，交易
delivery	n.	递送，交付
	n.	发送，传输
flexibility	n.	弹性，适应性，机动性
heighten	v.	提高，升高
inevitable	adj.	不可避免的，必然的
inquiry	n.	质询，调查
marketplace	n.	集会场所，市场，商场
oppose	v.	反对
pocketbook	n.	笔记本，钱袋，皮夹
procedure	n.	程序，手续
reduction	n.	减少，缩影，变形，缩减量，约简
restriction	n.	限制，约束
strengthen	v.	加强，巩固
transaction	n.	办理，交易，事务，处理事务
via	prep.	经，通过，经由

3.3.2 正文参考译文

电子商务的定义和内容

通过Internet进行的电子商务是一个新概念。近年来，这一概念已被广泛使用，几乎无异于当前依靠自动化的其他趋势(如工程学和即时生产)。很多公司(包括 CyberCash、DigCash、First Virtual 和 Open Market)提供了多种多样的电子商务服务。

如果你有机会使用一台个人计算机并且能连接上 Internet,你就能在线做生意。不用再为编程担心;客户再也不用寻找过时的目录,商家再也不用打印目录;再也不用查询电话号码、支付长途费用或者很晚还开着商店。只是上网,开一家网上商店,然后坐看你的生意增长。

发达技术、人的才能和新经营方式的网络世界构成今天日益增长的全球经济。电子商务的核心是 Internet。网络世界不是关于技术的,它是关于信息、决策和交流的。这个世界正在改变每个人的生活——从单个的家庭到大的公司,没有企业能忽视互联网经济的潜能。

作为新兴概念,电子商务描述了通过计算机网络(包括 Internet)买卖或交换商品、服务与信息的过程。Kalakota 和 Whinston(1997)从以下角度定义了电子商务:

从通信角度看,电子商务通过电话线路、计算机网络或其他电子方式传递信息、产品、服务,进行支付。

从商务角度看,电子商务是促使商务交易及工作流程自动化的技术应用。

从服务角度看,电子商务是满足公司、消费者和管理者的减少服务成本、提高产品质量、加快服务速度等愿望的工具。

从在线角度看,电子商务通过 Internet 及其联机服务提供购买和销售商品与信息的能力。

人们将商务看作商业伙伴之间的交易行为。因此,对于某些人来说,电子商务的概念有些偏窄。所以许多人使用 e-business 一词,即广义的电子商务。广义的电子商务不仅涉及到买与卖,还包括对客户的服务、商业伙伴之间的合作以及企业组织内部的电子交易。根据 IBM 公司首席执行官 Lou Gerstner 所言:"广义的电子商务是一个完整的周期,意味着高速度、全球化、增加产量、获得新的客户、共享竞争优势。"

正如任何其他类型商业一样,电子商务主要包含两个主体:企业和消费者。基本的电子商务类型有三种。

企业对消费者(B2C):这是一种面向个体购物者的零售交易。例如,亚马逊网站的客户就是典型的个体购物者。通常这种模式将产品直接销售给客户,免去了许多中间环节。另外,零售商通过网上经营作为联系客户的另一种方式。

消费者对消费者(C2C):在这一类型中,消费者与消费者之间直接进行交易,它们经常采用分类广告或拍卖的电子形式。货物在网上展示,感兴趣的买方联系卖方议价。与通过分类广告和拍卖方式的传统销售不同,买方与卖方通常从未见过面。例如,个人在分类广告中销售房产或轿车等。在 Internet 上为个人服务做广告、提供专业咨询服务是 C2C 的另外一个例证。有几个拍卖网站提供个人拍卖服务项目,还有许多人通过内联网提供销售或服务的广告信息。

企业对企业(B2B):这一类型涉及企业与企业之间进行产品和服务的交易,二者通常是一种制造商与供应者的关系。例如,一个家具制造商需要原料如木头、颜料以及清漆,在企业对企业电子商务过程中,制造商就可以以电子方式向供应商订购,款项也多以电子方式支付。

很多人认为电子商务就是有一个网站,但是电子商务远不止于此。电子商务的应用很广,如家庭银行业务、在线商店和商业区的购物、买股票、找工作、进行拍卖以及在网上合作研发项目等。为了执行这些应用，电子商务需要拥有支持信息和机构化的基础结构和系统。如图 3.1 所示基础设施支持电子商务应用，电子商务应用的实施依赖于四个主要方面(图示为支柱)：人、国家政策、技术标准和协议以及其他组织。电子商务管理者使应用、基础设施和支柱各部分协调工作。

3.3.3 阅读材料参考译文

电子商务及其优点

无论是否在线，今天的客户需要高质量的产品和快速简便的信息。互联网的主要好处是快速和便利。因此，电子商务利用互联网为核心进行业务往来，可以帮助公司在更加高效地交换信息/或产品以及服务等之外更方便顾客。

1. 超越位置和可用性的限制

用户无需在同一个物理位置，因为电子商务和信息交流和交易可以发生在任何时间(每天 24 小时，每个星期 7 天)和世界上可连接互联网的任何地点。实际位置受距离的限制，只限于服务那些可以到达的客户，而网上商店有个全球性的市场，客户和搜寻信息者遍布全世界。

2. 节约时间和金钱开支

电子商务中，完成传统业务的程序所需的费用往往会减少。电子手段可取代传统业务的许多方法，而且电子手段往往更容易开展，费用也更便宜。例如，发送 100 个普通邮件的费用(纸张、邮费、工作人员等)比发送大批量电子邮件的费用高；去一个实际位置支付房租的费用比维护联机站点的费用高。

3. 提升客户服务

使用电子商务，客户可收到高度自定义的服务，沟通往往更为有效。在线支持带来了更大的灵活性、可用性和响应速度。通过电子邮件查询和即时聊天的速度比使用电话通话方便多了，尤其是当白天的业务已经结束时。另外，在线销售缩短了交付周期，有助于加强客户/业务关系。互联网是获得新市场、对客户和合作伙伴传播信息的强效渠道，因为更好地了解客户将有助于提高客户满意度。

4. 具有竞争优势

互联网开辟了企业网上运行的全新市场。互联网在成长，通过互联网的竞争也日益增强，等待太久而耽误网上运行，可能会使你在竞争队伍中完全失去原来的位置。互联网的主要好处是很容易获得即时信息，能为公司提供更有效、有用的信息，帮助公司获得超过那些不上网的公司的竞争优势。

像大多数的业务决策一样，E-Business 也有相关风险。但是，无力适应时代变化更有风险。变化在当今市场不可避免，我们应该张开双臂、开放网站迎接时代变化！

3.4 Value Chains in E-commerce

[1]In e-commerce, a number of business process and activities go unnoticed by the consumer

and are often taken for granted. With an online merchant's business, value-added activities work together to make the business-to-consumer interface operational.

In 1985 Michael Porter wrote a book called Competitive Advantage, in which he introduced the concept of the value chain. Businesses receive raw materials as input, add value to them through various processes, and sell the finished product as output to customers.

Competitive advantage is achieved when an organization links the activities in its value chain more cheaply and more effectively than its competitors. For example, the purchasing function assists the production activity to ensure that raw materials and other supplies are available on time and meet the requirements of the products to be manufactured. The manufacturing function, in turn, has the responsibility to produce quality products that the sales staff can depend on. The human resource function must hire, retain, and develop the right personnel to ensure continuity in manufacturing, sale, and other areas of the business. Bringing in qualified people contributes to stability, continuity, and integrity of operations throughout the firm.

There is no time sequence or special sequence of activities before a business is considered successful or effective. The idea is to link different activities in such a way that the value-added (output) of one activity (department, process, etc.) contributes to the input of another activity. The integration of these activities results in an organization fine-tuned for profitability and growth.

According to Porter, the primary activities of a business are:

Inbound logistics. These are procurement activities vendor selection, comparative shopping, negotiating supply contracts, and just-in-time arrival of goods. They represent the supply side of the business. In e-commerce, the business must be capable of exchanging data with suppliers quickly, regardless of the electronic format.

Operations. This is the actual conversion of raw materials received into finished products. It includes fabrication, assembly, testing, and packaging the product. This production activity provides added value for the marketing function. Operational activities are the point in the value chain where the value is added. These happen in the back-office where the pizza is baked, the PCs are assembled, or the stock trades are executed. Data are shared at maximum network speed among internal and external partners involved in the value-adding processes.

Outbound logistics. This activity represents the actual storing, distributing, and shipping of the final product. It involves warehousing, materials handling, shipping, and timely delivery to the ultimate retailer or customer. The output of this activity ties in directly with marketing and sales.

Marketing and sales. This activity deals with the ultimate customer. It includes advertising, product promotion, sales management, identifying the product's customer base, and distribution channels. The output of this activity could trigger increased production, more advertising, etc.

Service. This activity focuses on after-sale service to the customer. It includes testing, maintenance, repairs, warranty work, and replacement parts. The output of this activity means satisfied customers, improved image of the product and the business, and potential for increased production, sale, etc.

Primary activities are not enough. A business unit needs support activities to make sure the primary activities are carried out. Imagine, for example, a manufacturing concerns with no people or with poorly skilled employees.

The key support activities in the value chain are:

Corporate infrastructure. This activity is the backbone of the business unit. It includes general management, accounting, finance, planning, and legal services. It is most often pictured in an organization chart showing the relationship among the different positions, the communication network, and the authority structure. Obviously, each position holder must add value to those above as well as below.

Human resources. This is the unique activity of matching the right people to the job. It involves recruitment, retention, career path development, compensation, training and development, and benefits administration. The output of this activity affects virtually every other activity in the company.

Technology development. This activity adds value in the way it improves the product and the business processes in the primary activities. The output of this activity contributes to the product quality, integrity, and reliability, which make life easier for the sales force and for customer relations.

Procurement. This activity focuses on the purchasing function and how well it ensures the availability of quality raw material for production.

Where does e-commerce fit in? The value chain is a useful way of looking at a corporation's activities and how the various activities add value to other activities and to the company in general.

[2]E-commerce can play a key role in reducing costs, improving product quality and integrity, promoting a loyal customer base, and creating a quick and efficient way of selling products and services. By examining the elements of the value chain, corporate executives can look at ways of incorporating information technology and telecommunications to improve the overall productivity of the firm. Companies that do their homework early and well ensure themselves a competitive advantage in the marketplace.

Words

administration	n.	管理，管理部门
analyze	v.	分析
assembly	n.	装配
bake	v.	烘，烤
chain	n.	链，表链
compensation	n.	补偿，赔偿，赔偿费
ensure	v.	确保

fabrication	n.	制作，构成
focus	n.	焦点
handling	n.	利用，管理，处理
inbound	adj.	回程的
incorporate	v.	结合，合并，收编
integration	n.	整体
logistic	n.	物流
loyal	adj.	忠诚的，忠心的
manufacture	v.	制造
overall	adj.	全面的
pizza	n.	比萨饼
procurement	n.	采购
raw	adj.	原始的
recruitment	n.	招募
retention	n.	保留，保持力
trigger	v.	触发，促使，引起
vendor	n.	卖主，小贩
warehousing	n.	仓库费，仓库储存
warranty	n.	担保，保证

Phrases

be taken for granted	认为某事理所当然
bring in	引进，带来
contribute to	有助于，促进
deal with	处理
deliver to	转交，交付，传达
fine-tune	调整，使有规则
play a key role	扮演一个重要的角色，起了一个重要的作用
value chains	价值链

Notes

[1] 例句：In e-commerce, a number of business process and activities go unnoticed by the consumer and are often taken for granted.

分析：本句是简单句。句中 go unnoticed 意思是"不被关注"，这里 go 是系动词，多表示"(从好的状态)变成坏的状态"。例如：In hot weather, meat goes bad. 热天，肉会变坏。I can't let this act of kindness go unnoticed. 我不能让这种好人好事湮没无闻。

译文：在电子商务中，消费者并不关注交易的进程和活动，并且经常认为那是理所当然的。

[2] 例句：E-commerce can play a key role in reducing costs, improving product quality and integrity, promoting a loyal customer base, and creating a quick and efficient way of selling products and services.

分析：本句是简单句，其中 reducing costs, improving product quality and integrity, promoting a loyal customer base, and creating a quick and efficient way of selling products and services 是并列的动名词短语，作 in 的宾语。

译文：电子商务能在降低成本、改善产品质量和完整性、促成忠诚客户基础和创建快捷高效的产品销售与服务方面起到关键作用。

Exercises

Ⅰ. Put "true" or "false" in the brackets for the following statements according to the passage.

1. (   ) With an online merchant's business, value-added activities work together to make the business-to-consumer interface operational.
2. (   ) The concept of the value chain is that the businesses receive finished product as input, and add value to them through various processes.
3. (   ) The purchasing function has the responsibility to produce quality products that the sales staff can depend on.
4. (   ) In e-commerce the business must be capable of exchanging data with suppliers quickly, regardless of the electronic format.
5. (   ) Operational activities are the point in the value chain where the value is added.
6. (   ) Outbound logistics represents the just-in-time arrival of goods.
7. (   ) The output of outbound logistics ties in directly with marketing and sales.
8. (   ) Marketing and sales focus on after-sale service to the customer.
9. (   ) The key support activities in the value chains are corporate infrastructure, human resources, technology development and procurement.
10. (   ) Technology development focuses on the purchasing function.

Ⅱ. Fill in the blanks according to the passage.

1. In e-commerce, a number of business process and activities go unnoticed by the consumer and are often _____.
2. Businesses receive _____ as input, add value to them through various processes, and sell the _____ as output to customers.
3. The human resource function must hire, retain, and develop the right personnel to ensure _____.
4. Bringing in qualified people contributes to _____, _____, _____ of operations throughout the firm.
5. In e-commerce, the business must be capable of _____ with suppliers quickly, regardless of the _____.
6. Outbound logistics represents the actual _____, _____ and _____ of the final product.

7. Corporate infrastructure is the _____ of the business unit.

8. Technology development adds value in the way it improves the _____ and _____ processes in the primary activities.

9. Procurement focuses on the _____ function and how well it ensures the availability of quality raw material for production.

10. E-commerce can play a key role in _____, _____ and _____, promoting a loyal customer base, and creating a quick and efficient way of selling products and services.

III. Translate the following words and expressions into Chinese.

1. Value Chains
2. fine-tuned
3. just-in time
4. inbound logistics
5. outbound logistics
6. after-sale service
7. backbone
8. human resources
9. purchasing function
10. product quality

3.4.1 Reading Material

ISP

An Internet Service Provider (ISP, also called Internet Access Provider or IAP) is a company that offers its customers access to the Internet. The ISP connects to its customers using a data transmission technology appropriate for delivering Internet Protocol Datagrams, such as dial-up, DSL, cable modem or dedicated high-speed interconnects.

ISPs may provide Internet e-mail accounts to users, which allow them to communicate with one another by sending and receiving electronic messages through their ISPs' servers. (As part of their e-mail service, ISPs usually offer the user an e-mail client software package, developed either internally or through an outside contract arrangement.) ISPs may provide other services such as remotely storing data files on behalf of their customers, as well as other services unique to each particular ISP.

End-User-to-ISP Connection

ISPs employ a range of technologies to enable consumers to connect to their network.

For home users and small businesses, the most popular options include dial-up, DSL (typically Asymmetric Digital Subscriber Line, ADSL), broadband wireless, cable modem, Fiber To The Premises (FTTP, Fiber To The Home FTTH), and Integrated Services Digital Network (ISDN) (typically basic rate interface).

For customers with more demanding requirements, such as medium-to-large businesses, or other ISPs, DSL (often SHDSL or ADSL), Ethernet, Metro Ethernet, Gigabit Ethernet, Frame Relay, ISDN (BRI or PRI), ATM, satellite Internet access and Synchronous Optical Networking (SONET) are more likely to be used.

Just as their customers pay them for Internet access, ISPs themselves pay upstream ISPs for

Internet access. An upstream ISP usually has a larger network than the contracting ISP and/or is able to provide the contracting ISP with access to parts of the Internet the contracting ISP by itself has no access to.

In the simplest case, a single connection is established to an upstream ISP and is used to transmit data to or from areas of the Internet beyond the home network; this mode of interconnection is often cascaded multiple times until reaching a Tier 1 carrier. In reality, the situation is often more complex. ISPs with more than one point of presence (PoP) may have separate connections to an upstream ISP at multiple PoPs, or they may be customers of multiple upstream ISPs and may have connections to each one of them at one or more point of presence.

A Virtual ISP (VISP) is an operation which purchases services from another ISP (sometimes called a "wholesale ISP" in this context) which allow the VISPs customers to access the Internet using services and infrastructure owned and operated by the wholesale ISP.

Free ISPs are Internet Service Providers (ISPs) which provide service free of charge. Many free ISPs display advertisements while the user is connected; like commercial television, in a sense they are selling the users' attention to the advertiser. Other free ISPs, often called freenets, are run on a nonprofit basis, usually with volunteer staff. There are also free shell providers and free web hosts.

Words

account	n.	户头，账目
appropriate	adj.	适当的
cascaded	n.	小瀑布，喷流，层叠
client	n.	[计]顾客，客户，委托人
contract	n.	合同，契约，婚约
	v.	使缩短，感染，订约
contracting	adj.	收缩的，缩成的，缩小的，缔约的
datagram	n.	[计]自带寻址信息的数据包
dedicated	adj.	专注的，献身的
delivering	v.	递送，陈述，释放，发表
dial-up	n.	刻度盘，转盘，(自动电话)拨号盘
	v.	拨号
establish	v.	建立，安置，使定居，确定
Ethernet	n.	以太网
interconnect	v.	(使)互相联系
internally	adv.	在内，在中心，在国内，在本国
nonprofit	adj.	非赢利的，不以赢利为目的的，无利可图的
purchase	v.	买，购买
	n.	买，购买
remotely	adv.	遥远地，偏僻地

staff	n.	棒，杆，支柱，全体职员
unique	adj.	唯一的，独特的
upstream	adv.	向上游，溯流，逆流地
	adj.	溯流而上的
volunteer	n.	志愿者，志愿兵
	adj.	志愿的，义务的，无偿的
wholesale	n.	批发，趸售
	adj.	批发的，[喻]大规模的

Phrases

broadband wireless	无线宽带
Fiber To The Home (FTTH)	光纤到户
Fiber To The Premises (FTTP)	光纤到楼宇
Frame Relay	帧中继
Gigabit Ethernet	千兆位以太网
on behalf of	作为……的代理，在……一边
unique to...	只有……才有的
Virtual ISP	虚拟的 Internet 服务提供者

Abbreviations

ADSL (Asymmetric Digital Subscriber Line)	非对称数字用户线
ATM (Asynchronous Transfer Mode)	异步传输模式
ISDN (Integrated Services Digital Network)	综合服务数字网
ISP (Internet Service Provider)	Internet 服务提供者

3.4.2 正文参考译文

电子商务中的价值链

在电子商务中，消费者并不关注交易的进程和活动，并且经常认为那是理所当然的。在线交易中，价值增值活动一起发生作用使得 B2C(企业对消费者)模式的交易得以实施。

1985 年迈克尔·波特写了《竞争优势》一书，在书中他介绍了价值链的概念。企业接受原材料为"输入"，经过各种不同的流程为其增加价值，然后将最终的产品作为"输出"销售给消费者。

当一个组织能比其他竞争者更有效更便宜地将价值链中的活动连接起来时，竞争优势就实现了。举例来说，采购部门辅助生产活动，确保原材料和其他供给及时到位并满足产品生产的需求。相应地，生产部门有责任生产出销售团队可依赖的高质量的产品。人力资源部门要雇佣、保留和发展合适的人才，以保证公司在制造、销售和其他环节上具有连续性。吸引高素质的人才才能保证整个公司的稳定持续发展。

一个交易在被认定是成功或有效之前，其行为没有时间顺序或特别顺序。这个想法是以下面的方式连接不同的商务活动：一个环节(部门、流程等)的增加价值(输出)直接作用于另一个环节的输入。这些环节的整合导致组织调整以获得收益和增长。

据波特介绍，商务活动(价值链的构成要素)主要包括：

内运物流。包含一些采购活动如供应商选择、比较购买、就合同的谈判和货物的即时抵达。它们代表着商务的供应方。在电子商务中，不管采用何种电子形式，企业必须有能力和供应商快速交换信息。

营运。营运是原材料转变成制成品的环节，包括制造、装配、测试和包装产品。这一生产活动增加了市场营销功能的价值。营运活动是价值链中增加价值的最重要环节，这些活动发生在后台，如比萨饼的烘烤、计算机的装配和股票的交易等。参与增值流程的公司内部人员和外部合伙人以最大的网络速度共享数据。

配送物流。该环节是存储、分配和运输最终产品的环节，包括仓储、材料处理、运输和及时发送到零售商或消费者。该环节的输出直接和市场销售相联系。

营销和销售。该环节与最终的消费者有关，包括广告、促销、销售管理、了解客户需求和销售渠道。该环节的输出可以提高产量，增加广告。

服务。该环节针对为客户提供售后服务，它包括测试、维护、修理、质量保证和退换产品。这个环节的输出是使客户满意，以提高产品和企业的形象，从而潜在地提高产销量。

有基本的商务价值活动并不够，一个企业需要支撑活动来确保基本商务活动的实施。设想一下，制造业没有人力资源或者说没有熟练员工能行吗？

价值链中的主要支撑活动是：

公司内部管理系统。公司的核心活动，包括一般的管理、会计、财政、计划和法制服务。在组织结构图中通常会展示不同位置之间的关系、联系网络和职权结构。很明显，每个位置的环节都会给其前后的环节增加价值。

人力资源。调配最优的人们到适合工作岗位的独特环节，包括招募、保持、职业发展、薪酬、培训与发展、津贴管理等。这个环节直接影响到公司其他所有环节。

技术研发。改进基本环节中的产品和业务流程的形式进行增值的活动。技术研发工作对产品质量、完整性和可信度有益，从而可以增强购买力，增进与顾客的关系。

采购。这个环节的主要功能是购买，确保得到高质量的原材料用于生产。

电子商务适合在什么环节呢？价值链是考虑一个公司如何运作的有效方式，利用价值链考虑各种不同的商务活动环节如何为其他商务活动环节增值、为整个公司增值。

电子商务能在降低成本、改善产品质量和完整性、促成忠诚客户基础和创建快捷高效的产品销售与服务方面起到关键作用。通过对价值链中的要素进行分析，公司管理人员可以考虑使用信息技术和远程通讯结合的各种方式来改进公司的总体生产力。那些尽早做好准备工作的公司能在市场中具备竞争优势。

3.4.3 阅读材料参考译文

Internet 服务提供商

Internet 服务提供商(ISP)，也称为 Internet 接入提供商(IAP)，是为客户提供接入互联网的公司。ISP 使用适当的数据传输技术连接到客户，传送互联网协议数据包。传输技术有拨号、DSL、电缆调制解调器或专用高速互连。

互联网服务供应商可以为用户提供互联网电子邮件账户，使它们通过 ISP 服务器发送和接收电子信息相互通信。(互联网服务供应商通常提供给用户一个电子邮件客户端软件,作为其电子邮件服务的一部分。该软件由供应商内部开发,或通过合同安排外部开发。)互联网服务供应商可提供其特有的服务,也可提供其他服务,如为其客户远程存储数据文件。

终端用户到 ISP 的连接

互联网服务供应商采用了一系列技术,使消费者能够连接到他们的网络。

对于家庭用户和小企业,最常用的技术包括拨号、DSL(通常是非对称数字用户线 ADSL)、宽带无线、电缆调制解调器、光纤到楼宇(光纤到户)、综合服务数字网(ISDN)(通常是基本速率接口)。

对于有更高要求的客户,如中型及大型企业或其他互联网服务供应商,则更可能采用 DSL(通常 SHDSL 或 ADSL)、以太网、城域以太网、千兆位以太网、帧中继、综合业务数字网(BRI 或 PRI)、ATM、卫星互联网接入和同步光纤网络(SONET)。

客户为接入互联网向 ISP 付费,ISP 也为接入互联网向上游供应商付费。与小的 ISP 相比,上游的 ISP 通常有较大的网络,能够提供给小的 ISP 没有接入的部分。

在最简单的情况下,与上游 ISP 建立一个单一的连接,用来与本地网络以外的互联网领域互传数据;这种互连模式往往级联多次,直至达到 1 级承运人。在现实中,情况则更为复杂。有多个接入点(POP)的互联网服务供应商可能在多个接入点有分别到上游 ISP 的连接,或者它们是多个上游供应商的客户,可能在一个或多个接入点连接到每个上游 ISP。

虚拟 ISP(VISP)则从另一个 ISP(这种情况下称其为"批发的 ISP")购买服务。批发的 ISP 允许虚拟 ISP 的客户使用其拥有和经营的服务以及基础设施来访问互联网。

免费 ISP 是提供免费服务的 ISP。许多免费的 ISP 在用户连接时展示广告;像商业电视台一样,从某种意义上他们出售用户的注意力给广告客户。其他免费 ISP(通常称为免费网络)运行在非营利性的基础上,通常有志愿工作人员,也有免费的壳供应商和免费的网络主机。

Chapter 4 Database Fundamentals

4.1 Introduction to DBMS

A database management system (DBMS) is an important type of programming system, used today on the biggest and the smallest computers. [1]As for other major forms of system software, such as compilers and operating systems, a well-understood set of principles for database management systems have been developed over the years, and these concepts are useful both for understanding how to use these systems effectively and for designing and implementing DBMS's. DBMS is a collection of programs that enables you to store, modify, and extract information from a database. There are many different types of DBMS's, ranging from small systems that run on personal computers to huge systems that run on mainframes.

DBMS Qualifies

There are two qualities that distinguish database management systems from other sorts of programming systems.

1) The ability to manage persistent data;
2) The ability to access large amounts of data efficiently.

Point 1) merely states that there is a database that exists permanently; the contents of this database are the data that a DBMS accesses and manages. Point 2) distinguishes a DBMS from a file system, which also manages persistent data. A DBMS's capabilities are needed most when the amount of data is very large, because for small amounts of data, simple access techniques, such as linear scans of the data, are usually adequate.

[2]While we regard the above two properties of a DBMS as fundamental, there are a number of other capabilities that are almost universally found in commercial DBMS's. They are:

- Support for at least one data model, or mathematical abstraction through which the user can view the data.
- Support for certain high-level languages that allow the user to define the structure of data, access data, and manipulate data.
- Transaction management, the capability to provide correct, concurrent access to the database by many users at once.
- Access control, the ability to limit access to data by unauthorized users, and the ability to check the validity of data.
- Resiliency, the ability to recover from system failures without losing data.

Data Models   Each DBMS provides at least one abstract model of data that allows the user to see information not as raw bits, but in more understandable terms. In fact, it is usually possible to see data at several levels of abstraction. At a relatively low level, a DBMS commonly allows us to visualize data as composed of files.

Efficient File Access The ability to store a file is not remarkable: the file system associated with any operating system does that. The capability of a DBMS is seen when we access the data of a file. For example, suppose we wish to find the manager of employee "Clark Kent". If the company has thousands of employees, It is very expensive to search the entire file to find the one with NAME="Clark Kent". A DBMS helps us to set up "index files", or "indices", that allow us to access the record for "Clark Kent" in essentially one stroke no matter how large the file is. Likewise, insertion of new records or deletion of old ones can be accomplished in time that is small and essentially constant, independent of the file's length. Another thing a DBMS helps us do is to navigate among files, that is, to combine values in two or more files to obtain the information we want.

Query Languages To make access to files easier, a DBMS provides a query language, or data manipulation language, to express operations on files. Query languages differ in the level of detail they require of the user, with systems based on the relational data model generally requiring less detail than languages based on other models.

Transaction Management Another important capability of a DBMS is the ability to manage simultaneously large numbers of transactions, which are procedures operating on the database. Some databases are so large that they can only be useful if they are operated upon simultaneously by many computers: often these computers are dispersed around the country or the world. The database systems used by banks, accessed almost instantaneously by hundreds or thousands of automated teller machines (ATM), as well as by an equal or greater number of employees in the bank branches, is typical of this sort of database. An airline reservation system is another good example.

Sometimes, two accesses do not interfere with each other. For example, any number of transactions can be reading your bank balance at the same time, without any inconsistency. [3]But if you are in the bank depositing your salary check at the exact instant your spouse is extracting money from an automatic teller, the result of the two transactions occurring simultaneously and without coordination is unpredictable. Thus, transactions that modify a data item must "lock out" other transactions trying to read or write that item at the same time. A DBMS must therefore provide some form of concurrency control to prevent uncoordinated access to the same data item by more than one transaction.

Even more complex problems occur when the database is distributed over many different computer systems, perhaps with duplication of data to allow both faster local access and to protect against the destruction of data if one computer crashes.

Security of Data A DBMS must not only protect against loss of data when crashes occur, as we just mentioned, but it must prevent unauthorized access. For example, only users with a certain clearance should have access to the salary field of an employee file.

DBMS Types

Designers developed three different types of database structures: hierarchical, network, and relational. Hierarchical and network were first developed but relational has become dominant. While the relational design is dominant, the older databases have not been dropped. Companies that installed a hierarchical system such as IMS in the 1970s will be using and maintaining these databases for years to come even though new development is being done on relational systems. These older systems are often referred to as legacy systems.

Words

compiler	n.	编译器
coordination	n.	同等，调和
concurrency	n.	同时(或同地)发生，同时存在，合作
clearance	n.	清除
distinguish	v.	区别，辨别
duplication	n.	副本，复制
destruction	n.	破坏，毁灭
fundamental	adj.	基础的，基本的
indices	n.	(index 的复数)索引，指针
instantaneously	adv.	瞬间地，即刻地，即时地
legacy	n.	遗赠(物)，遗产
manipulate	v.	(熟练的)操作，使用(机器等)
navigate	v.	导航，航行，航海，航空
permanently	adv.	永存地，不变地
resiliency	n.	跳回，弹性
reservation	n.	保留，(旅馆房间等的)预定，预约
transaction	n.	办理，学报，交易，处理事务
unauthorized	adj.	未被授权的，未经认可的

Phrases

at the exact instant	此时，与此同时，正当此时
associate with	联合，结合
at will	随意，任意
high-level languages	高级语言
take into account	重视，考虑

Abbreviations

ATM(Automatic Teller Machine)	自动取款(出纳)机
DBMS(Database Management System)	数据库管理系统

Chapter 4　Database Fundamentals

Notes

[1] 例句：As for other major forms of system software, such as compilers and operating systems, a well-understood set of principles for database management systems have been developed over the years, and these concepts are useful both for understanding how to use these systems effectively and for designing and implementing DBMS's.

分析：本句是一个并列句。句首的 As for 用来转换话题，起到补充前文又突出后文的作用，可译为：至于。例句：As for the future of garments, I think fashion will become garish .至于未来的衣服，我认为时尚会变得过分装饰。句中的 such as…systems 之间的部分是插入语，翻译时可以把两个逗号作为括号翻译出来。

译文：至于其他主要形式的系统软件(如编译器以及操作系统)，近些年来开发出一系列容易理解的数据库管理系统原则，这些概念既有助于理解如何有效利用系统，又可以帮助设计和实现 DBMS 系统。

[2] 例句：While we regard the above two properties of a DBMS as fundamental, there are a number of other capabilities that are almost universally found in commercial DBMS's.

分析：本句是一个复合句，while 放于句首引导让步状语从句，翻译为"虽然"，后半句转折，英文中不用"but"，但在翻译时需要翻译出"虽然……，但是……"。

译文：虽然我们将以上两点作为 DBMS 的基本特性，但是其他一些功能在商业 DBMS 系统中也是常见的。

[3] 例句：But if you are in the bank depositing your salary check at the exact instant your spouse is extracting money from an automatic teller, the result of the two transactions occurring simultaneously and without coordination is unpredictable.

分析：本句是一个复合句，主句为"the result of…unpredictable"，前面部分为 if 所引导的表假设的条件从句。条件从句中的 your spouse is extracting money from an automatic teller 是定语从句，修饰 the exact instant。

译文：但是如果你正在银行里办理工资存款，与此同时，你的配偶在一台自动取款机上取款，两个事务同时发生且没有彼此协调，那你的查询结果就很难说了。

Exercises

Ⅰ. Put "true" or "false" in the brackets for the following statements according to the passage.

1. (　) A database management system is an important type of programming system.
2. (　) There is only one type of DBMS.
3. (　) DBMS has the ability to access large amounts of data efficiently.

4. (　) At a relatively low level, a DBMS commonly hasn't the ability to allows us to visualize data as composed of files.

5. (　) A DBMS provides a query language to make access to files.

6. (　) Not all types of DBMS have the ability to manage simultaneously large numbers of transactions.

7. (　) DBMS can't prevent unauthorized access.

8. (　) A DBMS must protect against loss of data when crashes occur.

9. (　) While the relational design is dominant, the older databases have been dropped.

10. (　) Designers have developed four different types of database structures.

II. Fill in the blanks according to the passage.

1. DBMS is a collection of programs that enables you to _____, _____, and information from a database.

2. There are many different types of DBMSs, ranging from small systems that run on _____ to huge systems that run on _____.

3. DBMS support for at least one data model, or _____ through which the user can view the data.

4. DBMS support for certain _____ that allow the user to define the structure of data, access data, and manipulate data.

5. _____, the capability of DBMS to provide correct, concurrent access to the database by many users at once.

6. _____, the ability of DBMS to limit access to data by unauthorized users, and the ability to check the validity of data.

7. To make access to files easier, a DBMS provides a _____, or data manipulation language, to express operations on files.

8. A DBMS must protect against _____ when crashes occur.

9. Designers developed three different types of database structures: _____, _____, and _____.

10. Older hierarchical systems are often referred to as _____.

III. Translate the following words and expressions into Chinese.

1. database management system
2. application program
3. end-user
4. file system
5. high-level language
6. query language
7. transaction management
8. relational data model
9. automatic teller machine
10. hierarchical database

4.1.1　Reading Material

A Database Administrator

A database administrator (DBA) is a person who is responsible for the environmental aspects of a database. The role of a database administrator has changed according to the technology of database management systems (DBMSs) as well as the needs of the owners of the databases. For example, although logical and physical database designs are traditionally the duties of a database analyst or database designer, a DBA may be tasked to perform those duties.

The duties of a database administrator vary and depend on the job description, corporate and Information Technology (IT) policies and the technical features and capabilities of the DBMS being administered. They nearly always include disaster recovery (backups and testing of backups), performance analysis and tuning, data dictionary maintenance, and some database design.

Some of the roles of the DBA may include:

◆ Installation of new software—It is primarily the job of the DBA to install new versions of DBMS software, application software, and other software related to DBMS administration. It is important that the DBA or other IS staff members test this new software before it is moved into a production environment.

◆ Configuration of hardware and software with the system administrator—In many cases the system software can only be accessed by the system administrator. In this case, the DBA must work closely with the system administrator to perform software installations, and to configure hardware and software so that it functions optimally with the DBMS.

◆ Security administration—One of the main duties of the DBA is to monitor and administer DBMS security. This involves adding and removing users, administering quotas, auditing, and checking for security problems.

◆ Data analysis—The DBA will frequently be called on to analyze the data stored in the database and to make recommendations relating to performance and efficiency of that data storage. This might relate to the more effective use of indexes, enabling "Parallel Query" execution, or other DBMS specific features.

◆ Database design (preliminary)—The DBA is often involved at the preliminary database-design stages. Through the involvement of the DBA, many problems that might occur can be eliminated. The DBA knows the DBMS and system, can point out potential problems, and can help the development team with special performance considerations.

◆ Data modeling and optimization—By modeling the data, it is possible to optimize the system layouts to take the most advantage of the I/O subsystem.

◆ Responsible for the administration of existing enterprise databases and the analysis, design, and creation of new databases.

- Data modeling, database optimization, understanding and implementation of schemas, and the ability to interpret and write complex Structured Query Language (SQL) queries
- Proactively monitor systems for optimum performance and capacity constraints
- Establish standards and best practices for SQL
- Interact with and coach developers in SQL scripting

Words

administrator	n.	管理人，行政官
analyst	n.	分析家，分解者
audit	n.	审计，稽核，查账
	v.	稽核，旁听，查账
capability	n.	(实际)能力，性能，容量，接受力
coach	n.	教练
	v.	训练，指导
configuration	n.	构造，结构，配置，外形
constraint	n.	约束，强制，局限
corporate	adj.	社团的，法人的，共同的，全体的
disaster	n.	灾难
efficiency	n.	效率，功效
environmental	adj.	周围的，环境的
execution	n.	实行，完成，执行
feature	n.	特征，容貌，特色，特写
index	n.	索引，指标
	v.	编入索引中，指出，做索引
involve	v.	包括，笼罩，使陷于
involvement	n.	连累，包含
layout	n.	规划，设计，版面，设计图案，版面设计
optimally	adv.	最适宜地，最理想地，最好地
optimization	n.	最佳化，最优化
optimum	n.	最适宜
	adj.	最适宜的
potential	adj.	潜在的，可能的
preliminary	adj.	预备的，初步的
proactive	adj.	[心理]前摄的
quota	n.	配额，限额
recommendation	n.	推荐，介绍(信)，劝告，建议
responsible	adj.	有责任的，可靠的，可依赖的，负责的
schema	n.	计划
staff	n.	全体职员
traditionally	adv.	传统上，传说上

Abbreviations

DBA(Database Administrator)	数据库管理员

4.1.2 正文参考译文

DBMS 简介

数据库管理系统是重要的一种编程系统，现今可以用在最大的以及最小的电脑上。至于其他主要形式的系统软件(如编译器以及操作系统)，近些年来开发出一系列容易理解的数据库管理系统原则，这些概念既有助于理解如何有效利用系统，又可以帮助设计和实现 DBMS 系统。DBMS 是一个程序的集合，它使你能够存储、修改以及从数据库中提取信息。DBMS 系统有很多不同的类型，从运行在个人电脑上的小型系统直到运行在大型主机上的巨型系统。

DBMS 的功能

有两种功能使数据库管理系统区别于其他程序设计系统：
1) 管理固有数据的能力；
2) 高效访问大量数据的能力。

第一点只是表明现有一个固定存在的数据库，该数据库的内容是 DBMS 要访问和管理的那些数据。第二点将 DBMS 和同样能管理固有数据的文件系统区分开来。在数据量非常大的时候最需要用到 DBMS 系统的功能，因为对于少量数据而言，简单的访问技术(如对数据的逐行扫描)就足够了。

虽然我们将以上两点作为 DBMS 的基本特性，但是其他一些功能在商业 DBMS 系统中也是常见的，它们是：
- 支持至少一种用户可以据之浏览数据的数据模式或数学提取方式。
- 支持某种允许用户用来定义数据的结构、访问和操纵数据的高级语言。
- 事务管理，即对多个用户提供正确地同时访问数据库的能力。
- 访问控制，即限制未被授权用户访问数据以及检测数据有效性的能力。
- 恢复功能，即能够从系统错误中恢复而不丢失数据的能力。

数据模型 每个 DBMS 提供了至少一种抽象数据模型，该模型允许用户以更容易理解的术语而不是以原始比特位的方式查看信息。实际上，通常可以做到以几个不同抽象级别观察数据。在相对低的级别中，DBMS 一般允许我们将数据形象化为由文件组成。

高效数据访问 存储一个文件的能力并不特别：操作系统中有的文件系统都能做到。DBMS 的能力在我们访问文件的数据时才能显示出来。比如，假设我们希望找到员工经理"克拉克·肯特"。如果这个公司有数千员工，要通过 NAME＝"克拉克·肯特"搜索整个文件来找到这个人肯定非常费时。而 DBMS 帮助我们建立"索引文件"或"索引"，不管文件有多大，它都使我们能够一举访问到"克拉克·肯特"的记录。同样的，不管文件的大小如何，新记录的插入或者原有记录的删除都可以在较短并且基本稳定的时间内完成。DBMS 还可以帮助我们进行文件间的导航，即通过结合两个或更多文件的值来获得我们所需的信息。

查询语言 为了使访问文件更容易，DBMS 提供了查询语言(或者说数据控制语言)表述对文件的操作。查询语言对用户所提供的细节的详细程度要求不同，基于关系数据模型的系统通常比基于其他模型的语言所需的细节少。

事务管理 DBMS 的另外一项重要功能就是同时管理大量事务的能力，事务即数据库中运行的进程。某些数据库非常大，它们只有同时运行于多台计算机上才能发挥作用：通常这些计算机分散在全国甚至世界各地。银行中使用的数据库系统就是这类数据库的一个典型，它们几乎同时被成千上万的自动取款机访问，也同时被同样多甚至更多的支行员工访问。机票预定系统是另一个好例子。

有时两个访问不会互相干扰。例如，任意多的事务可以同时读取你银行的结余而不引起任何冲突。但是如果你正在银行里办理工资存款，与此同时，你的配偶在一台自动取款机上取款，两个事务同时发生且没有彼此协调，那你的查询结果就很难说了。因此，会引起数据项改变的事务必须"上锁"，将其他在同一时刻试图读写该项数据的事务关在外面。因此，DBMS 必须提供某种并发控制，以阻止多个事务对于同一数据项的非协调访问。

更复杂的问题发生在数据库分布在许多不同计算机系统上的时候，它们可能使用数据副本来允许高速的本地访问，避免由于某台计算机崩溃而破坏数据。

数据安全 如上文提到的那样，DBMS 不仅可以在计算机崩溃时保护数据不丢失，而且还能够阻止非法访问。比如，只有拥有特定权限的用户可以访问职工文件的工资区域。

DBMS 的类型

设计人员开发了三种不同类型的数据库结构：层次数据库、网状数据库以及关系数据库。层次数据库和网状数据库是首先被开发出来的，但关系数据库已经成为了主流数据模型。尽管关系数据库的设计已经成为主流，但旧的数据库仍然没有被遗弃。尽管关系数据库不断得到发展，但在 20 世纪 70 年代安装了层次数据库的一些公司，如 IMS，在未来几年仍然将继续使用这些数据库。这些旧的数据库系统通常被称作遗留系统。

4.1.3 阅读材料参考译文

数据库管理员

数据库管理员(DBA)负责一个数据库环境的方方面面。数据库管理员的角色根据数据库管理系统的技术以及数据库业主的需要已经改变。例如，虽然逻辑和物理的数据库设计在传统上是数据库分析师或数据库设计师的职责，而现在数据库管理员可能会负责执行这些职责。

数据库管理员的职务各不相同，取决于工作说明、企业和信息技术(IT)的政策、技术特性以及被管理的数据库管理系统的功能。它们几乎总是包括灾难恢复(备份和测试备份)、性能分析和优化、数据字典维护以及一些数据库设计。

数据库管理员的职责包括：

◆ 安装新的软件——这是数据库管理员的主要工作，包括安装新版本的数据库管理系统软件、应用软件以及其他与数据库管理系统管理相关的软件。重要的是，数据库管理员或其他信息系统工作人员，在进入生产环境前需要先测试这个新的软件。

◆ 与系统管理员一起配置硬件和软件——在许多情况下，系统管理员才能访问系统软件。这样，数据库管理员和系统管理员必须紧密合作进行软件安装、配置硬件和软件，以便系统与数据库管理系统功能达到最优。

◆ 安全管理——数据库管理员的一个主要职责是监测和管理数据库管理系统的安

全。这包括添加和删除用户、管理配额、审计和检查安全问题。
- ◆ 数据分析——数据库管理员应经常分析存储在数据库中的数据，并提出有关数据存储的性能和效率的建议。这可能涉及到更有效地利用索引实现"并行查询"的执行，或涉及到其他数据库管理系统的具体作用。
- ◆ 数据库设计(初步)——数据库管理员往往参与数据库的初步设计阶段。通过数据库管理员的参与，许多可能发生的问题可以消除。数据库管理员了解该数据库管理系统和整个系统，指出潜在的问题，并能帮助开发团队考虑特殊性能。
- ◆ 数据建模和优化——通过数据建模，可以优化系统的布局，控掘 I/O 子系统的最大优势。
- ◆ 负责管理现有的企业数据库，分析、设计、建立新的数据库。
- 数据建模，数据库优化，理解和执行架构，并能够解释和编写复杂的结构化查询语言(SQL)查询。
- 积极主动地监控系统，以实现最佳的性能和容量限制。
- 为 SQL 建立标准和最佳准则。
- 和开发商配合并指导编写 SQL 脚本。

4.2　Structure of the Relational Database I

The relational model is the basis for any relational database management system (RDBMS). [1]A relational model has three core components: a collection of objects or relations, operators that act on the objects or relations, and data integrity methods. In other words, it has a place to store the data, a way to create and retrieve the data, and a way to make sure that the data is logically consistent.

A relational database uses relations, or two-dimensional tables, to store the information needed to support a business. Let's go over the basic components of a traditional relational database system and look at how a relational database is designed. Once you have a solid understanding of what rows, columns, tables, and relationships are, you'll be well on your way to leveraging the power of a relational database.

Tables, Rows, and Columns

A table in a relational database, alternatively known as a relation, is a two-dimensional structure used to hold related information. A database consists of one or more related tables.

Note：Don't confuse a relation with relationships. A relation is essentially a table, and a relationship is a way to correlate, join, or associate two tables.

A row in a table is a collection or instance of one thing, such as one employee or one line item on an invoice. [2]A column contains all the information of a single type, and the piece of data at the intersection of a row and a column, a field, is the smallest piece of information that can be retrieved with the database's query language. For example, a table with information about employees might have a column called LAST_NAME that contains all of the employees' last names. Data is retrieved from a table by filtering on both the row and the column.

Primary Keys, Datatypes, and Foreign Keys

The examples throughout this article will focus on the hypothetical work of Scott Smith, database developer and entrepreneur. He just started a new widget company and wants to implement a few of the basic business functions using the relational database to manage his Human Resources (HR) department.

Relation: A two-dimensional structure used to hold related information, also known as a table.

Note: Most of Scott's employees were hired away from one of his previous employers, some of whom have over 20 years of experience in the field. As a hiring incentive, Scott has agreed to keep the new employees' original hire date in the new database.

Row: A group of one or more data elements in a database table that describes a person, place, or thing.

Column: The component of a database table that contains all of the data of the same name and type across all rows.

You'll learn about database design in the following sections, but let's assume for the moment that the majority of the database design is completed and some tables need to be implemented. Scott creates the EMP table to hold the basic employee information, and it looks something like this:

EMPNO	ENAME	JOB	MGR	HIREDATE	SAL	COMM	DEPTNO
7369	SMITH	CLERK	7902	17-DEC-80	800		20
7499	ALLEN	SALESMAN	7698	20-FEB-81	1600	300	30
7521	WARD	SALESMAN	7698	22-FEB-81	1250	500	30
7566	JONES	MANAGER	7839	02-APR-81	2975		20
7839	KING	PRESIDENT		17-NOV-81	5000		10
7902	FORD	ANALYST	7566	03-DEC-81	3000		20

Notice that some fields in the Commission (COMM) and Manager (MGR) columns do not contain a value; they are blank. A relational database can enforce the rule that fields in a column may or may not be empty. [3]In this case, it makes sense for an employee who is not in the Sales department to have a blank Commission field. It also makes sense for the president of the company to have a blank Manager field, since that employee doesn't report to anyone.

Field: The smallest piece of information that can be retrieved by the database query language. A field is found at the intersection of a row and a column in a database table.

On the other hand, none of the fields in the Employee Number (EMPNO) column are blank. The company always wants to assign an employee number to an employee, and that number must be different for each employee. One of the features of a relational database is that it can ensure that a value is entered into this column and that it is unique. The EMPNO column, in this case, is the primary key of the table.

Primary Key: A column (or columns) in a table that makes the row in the table distinguishable from every other row in the same table.

Notice the different datatypes that are stored in the EMP table: numeric values, character or alphabetic values, and date values.

As you might suspect, the DEPTNO column contains the department number for the employee. But how do you know what department name is associated with what number? Scott created the DEPT table to hold the descriptions for the department codes in the EMP table.

DEPTNO	DNAME	LOC
10	ACCOUNTING	NEW YORK
20	RESEARCH	DALLAS
30	SALES	CHICAGO
40	OPERATIONS	BOSTON

The DEPTNO column in the EMP table contains the same values as the DEPTNO column in the DEPT table. In this case, the DEPTNO column in the EMP table is considered a foreign key to the same column in the DEPT table.

A foreign key enforces the concept of referential integrity in a relational database. The concept of referential integrity not only prevents an invalid department number from being inserted into the EMP table, but it also prevents a row in the DEPT table from being deleted if there are employees still assigned to that department.

Foreign Key: A column (or columns) in a table that draws its values from a primary or unique key column in another table. A foreign key assists in ensuring the data integrity of a table.

Referential Integrity: A method employed by a relational database system that enforces one-to-many relationships between tables.

Words

alphabetic	*adj.*	字母的
assign	*v.*	分配，指派
dimensional	*adj.*	空间的，[数]……维的
hypothetical	*adj.*	假设的
incentive	*n.*	动机
intersection	*n.*	[数]交集，十字路口，交叉点
query	*n.*	查询
relational	*adj.*	相关的

| suspect | v. | 猜想，怀疑 |
| widget | n. | 装饰品，小器具 |

Phrases

line item	排列项，项目
foreign key	外键
primary key	主键

Notes

[1] 例句：A relational model has three core components: a collection of objects or relations, operators that act on the objects or relations, and data integrity methods.

分析：本句为复合句。其中冒号后面部分为 three components 的同位语，collection 数学中意为"集合"。

译文：一个关系模型有三个核心组件：对象或关系的集合、作用于对象或关系上的操作符以及数据完整性规则。

[2] 例句：A column contains all the information of a single type, and the piece of data at the intersection of a row and a column, a field, is the smallest piece of information that can be retrieved with the database's query language.

分析：本句是并列句。在第二个分句中，"a field"是 the piece of data at the intersection of a row and a column 的同位语。"field"在数据库中指的是"字段"，而"database's query language"指的是数据库查询语言。

译文：表中的一列包含了一类信息；行列交叉点上的数据——字段，是能够用数据库查询语言检索到的最小信息片段。

[3] 例句：In this case, it makes sense for an employee who is not in the Sales department to have a blank Commission field.

分析：本句是一个简单句。其中，"make sense"意思是：讲得通，有意义。

译文：在这种情况下，非销售部门的员工拥有空白的销售提成字段是讲得通的。

Exercises

Ⅰ. Put "true" or "false" in the brackets for the following statements according to the passage.

1. (　) A relational database uses two-dimensional tables to store the information needed to support a business.

2. (　) A relation is the same as a relationship.

3. (　) A database consists of one or more related tables.

4. (　) A column in a table is a collection of one thing.

5. (　) A row in a table contains all the information of a single type.

6. (　) A primary key is a column (or columns) in a table that makes the row in this table distinguishable from every other row in every other table in the same database.

7. (　) A relation is essentially a database, and a relationship is a table in the database.

Chapter 4 Database Fundamentals

8. ( ) A field is the smallest piece of information that can be retrieved by the database query language.

9. ( ) A foreign key is a column (or columns) in a table that draws its values from a primary or unique key column in another table.

10. ( ) The referential integrity is a method employed by a relational database system that enforces one-to-many relationships between tables.

II. Fill in the blanks according to the passage.

1. The _____ is the basis for any relational database management system (RDBMS).

2. A relational model has three core components: _____ , _____ and_____ .

3. A _____ is a way to correlate, join, or associate two tables.

4. A relational database can enforce the rule that _____ may or may not be empty.

5. A Field is the smallest piece of information that can be retrieved by the _____ .

6. A field is found at the _____ of a row and a column in a database table.

7. A relational database uses relations, or _____, to store the information needed to support a business.

8. A _____ in a table is a collection or instance of one thing, such as one employee or one line item on an invoice.

9. A foreign key assists in ensuring the _____ of a table.

10. _____ is a method employed by a relational database system that enforces one-to-many relationships between tables.

III. Translate the following words and expressions into Chinese.

1. core components
2. two-dimensional table
3. related table
4. Foreign Key
5. Primary Key
6. numeric value
7. Referential Integrity
8. row
9. column
10. field

4.2.1 Reading Material

Database Design

A database must be carefully designed from the outset if it is going to fulfill its role. Poor planning in the early stages can lead to a database that does not support the necessary relationships that are required.

One popular technique for designing relational databases is called Entity-Relationship (ER) modeling. Chief among the tools used for ER modeling is the ER diagram. An ER diagram captures the important record types, attributes, and relationships in a graphical form. From an ER diagram, a database manager can define the necessary schema and create the appropriate tables to support the database specified by the diagram.

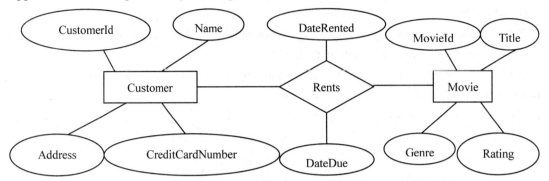

Figure 4.1 An ER diagram for the movie rental database

An ER diagram showing various aspects of the movie rental example is shown in Figure 4.1. Specific shapes are used in ER diagrams to differentiate among the various parts of the database. Types of records (which can also be thought of as classes for the database objects) are shown in rectangles. Fields (or attributes) of those records are shown in attached ovals. Relationships are shown in diamonds.

The position of the various elements of an ER diagram is not particularly important, though if some thought is given to it they are easier to read. Note that a relationship such as Rents can have its own associated attributes.

Also note that the relationship connectors are labeled, one side with a 1 and the other side with an M. These designations show the cardinality constraint of the relationship. A cardinality constraint puts restrictions on the number of relationships that may exist at one time.

There are three general cardinality relationships:

- one-to-one
- one-to-many
- many-to-many

The relationship between a customer and a movie is one-to-many. That is, one customer is allowed to rent many movies, but a movie can only be rented by a single customer (at any given time). Cardinality constraints help the database designer convey the details of a relationship.

Words

aspect	n.	样子，外表，(问题等的)方面
associated	adj.	联合的，关联的
capture	v.	捕获，记录，以影片、文字等保存原状
cardinality	n.	基数性
constraint	n.	约束，强制，局促
convey	v.	搬运，传达，转让
designation	n.	指示，指定，名称
diagram	n.	图表，图解

oval	n.	椭圆形
rectangle	n.	长方形，矩形
rental	n.	租金额，租贷
	adj.	租用的

4.2.2 正文参考译文

关系数据库的结构（一）

关系模型是所有关系数据库管理系统(RDBMS)的基础。一个关系模型有三个核心组件：对象或关系的集合、作用于对象或关系上的操作符以及数据完整性规则。换句话说，关系数据库有一个存储数据的地方、一种创建和检索数据的方法、以及一种确认数据的逻辑一致性的方法。

一个关系数据库使用关系或二维表来存储支持某个事物所需的信息。让我们了解一下传统的关系数据库系统的基本组件，并且学习如何设计关系数据库。对行、列、表和关联有了深刻理解后，就能够充分利用关系数据库的强大功能。

表、行和列

在关系数据库中，表(或者说关系)是用于保存相关信息的二维结构。一个数据库由一个或者多个相关联的表组成。

注意：不要混淆了关系和关联。一个关系实际上是一个表，而关联指的是一种连接、结合或联合两个表的方式。

表中的一行是一种事物的集合或实例，比如一个员工或发票上的一项。表中的一列包含了一类信息：行列交叉点上的数据——字段，是能够用数据库查询语言检索到的最小信息片段。例如，一个员工信息表可能有一个"姓名"列，列中就包含所有员工的姓名。数据是通过对行、列进行过滤而从表中检索出来的。

主键、数据类型和外键

接下来以假设的斯科特·史密斯的工厂为例来说明，史密斯是数据库的建立者和企业的主办人。他刚开办了一家饰品公司，想要使用关系数据库的几项基本功能来管理人力资源部门。

关系：用来保存相关信息的一个二维结构(也就是表)。

注意：斯科特的大多数雇员都曾在斯科特之前的一个老板手下工作，他们中有些人在这个领域已经有 20 多年的经验。作为雇佣的鼓励，斯科特同意在新数据库中员工的雇佣日期仍最初的雇佣日期。

行：数据库表中的一组单数据或多数据元素，用于描述人、地方或事物。

列：列是数据库表的组件，它包含所有行中同名和同类型的所有数据。

你会在下面章节学到如何设计数据库。现在让我们假设数据库大部分已经设计完成，有一些表需要填写。斯科特创建了 EMP 表来保存基本的员工信息，就像这样：

员工编号	名字	工种	管理人	工作日期	工资	佣金	部门成员
7369	史密斯	会计	7902	17-12-80	800		20
7499	艾伦	推销员	7566	20-2-81	1600	300	30
7521	沃德	推销员	7566	22-2-81	1250	500	30
7566	琼斯	管理人员	7839	02-4-81	2975		20
7839	金	厂长		17-11-81	5000		10
7902	福特	分析师	7566	03-12-81	3000		20

注意佣金列和管理人列中有一些单元格中没有值，它们是空值。一个关系数据库能够规定列中的一个单元格是否为空，因为非销售部门的员工可能拥有空白的销售提成字段。公司董事长的管理人单元为空也是可以的，因为董事长不需要向任何人汇报工作。

字段：是数据库查询语言所能够检索到的最小信息片段。字段位于数据库表的行和列的交叉处。

另一方面，没有哪个员工的员工编号单元为空。公司总是希望为每个员工分配一个员工号，并且这个号码必须是每个员工都不同。关系数据库的一个特性是能够确定该列必须有输入值且值唯一。这样，员工编号列便是这个表的主键。

主键：主键即是表中的一列(或多列)，它使得每一行能够区别于同一个表中的其他行。

EMP 表中存储的不同数据类型，可以是：数值型、字符型或字母型以及日期型。

如你所想，部门成员列保存的是员工所在部门的编号。但是如何知道哪个部门名称对应哪个部门编号呢？斯科特建立了 DEPT 表来具体描述 EMP 表中的部门编号。

部门编号	部门名称	位置
10	会计部	纽约
20	调查部	达拉斯
30	销售部	芝加哥
40	业务部	波士顿

EMP 表中的部门编号列同 DEPT 表中的部门编号列有着相同的值。这样 EMP 表中的部门编号列便被看作是与 DEPT 表中相同列对应的外键。

外键加强了关系数据库中参考完整性的概念。参考完整性的概念不只可以阻止无效的部门编号被插入 EMP 表中，而且可以防止 DEPT 表中有员工的部门的信息被删除。

外键：表中的一列(或多列)，它的值来自于其他表的主键列或唯一值列。外键有助于保障表中数据的完整性。

参考完整性：是关系数据库用来加强表间一对多关联的一种方式。

4.2.3 阅读材料参考译文

数据库设计

一个数据库从开始就要精心设计才能够较好地发挥作用。最初的计划不周可能导致一个数据库不支持所需要的关系。

一个设计关系数据库的常用技术被称为实体关系(ER)建模，ER 建模的主要工具是 ER 图。ER 图以图表形式记录重要的记录类型、属性以及关系。从 ER 图，数据库设计者可以定义必须的计划，并建成适当的表格以支持 ER 图描述的数据库。

图 4.1 的 ER 图，从不同角度展示影片出租实例。在 ER 图中采用特殊形状来区分数据库的不同部分。记录类型(也可以认为是数据库对象的类)用矩形表示。记录的字段(或属性)用的椭圆表示。关联用菱形表示。

即使位置会使得元素更容易读，ER 图的各种元素的位置也不特别重要。注意像 Rents 这样的关联也有它自己相关的属性。

还要注意关系连接要有标注：一边是 1，另一边是 M。这些标注表示关联的基数约束。一个基数约束限制同时存在的关联数。

常见的约束关联有三个：
- 一对一
- 一对多
- 多对多

客户和影片的关联是一对多关联。也就是说，一个客户可以租借多部影片，但是一部影片只能借给一个客户(在任何给定时间)。基数约束帮助数据库设计者表达关联的细节。

4.3　Structure of the Relational Database II

Data Modeling

Before Scott created the actual tables in the database, he went through a design process known as *data modeling*. In this process, the developer conceptualizes and documents all the tables for the database. One of the common methods for modeling a database is called **ERA**, which stands for entities, relationships, and attributes. The database designer uses an application that can maintain entities, their attributes, and their relationships. In general, an entity corresponds to a table in the database, and the attributes of the entity correspond to columns of the table.

Data Modeling：A process of defining the entities, attributes, and relationships between the entities in preparation for creating the physical database.

The data-modeling process involves defining the entities, defining the relationships between those entities, and then defining the attributes for each of the entities. Once a cycle is complete, it is repeated as many times as necessary to ensure that the designer is capturing what is important enough to go into the database. Let's take a closer look at each step in the data-modeling process.

Defining the Entities

First, the designer identifies all of the entities within the scope of the database application. The entities are the persons, places, or things that are important to the organization and need to be tracked in the database. Entities will most likely translate neatly to database tables. For example, for the first version of Scott's widget company database, he identifies four entities: employees, departments, salary grades, and bonuses. These will become the EMP, DEPT, SALGRADE, and

BONUS tables.

Defining the Relationships between Entities

Once the entities are defined, the designer can proceed with defining how each of the entities is related. Often, the designer will pair each entity with every other entity and ask, "Is there a relationship between these two entities?" Some relationships are obvious; some are not.

[1]In the widget company database, there is most likely a relationship between EMP and DEPT, but depending on the business rules, it is unlikely that the DEPT and SALGRADE entities are related. If the business rules were to restrict certain salary grades to certain departments, there would most likely be a new entity that defines the relationship between salary grades and departments. This entity would be known as an associative or intersection table and would contain the valid combinations of salary grades and departments.

Associative Table： A database table that stores the valid combinations of rows from two other tables and usually enforces a business rule. An associative table resolves a many-to-many relationship.

In general, there are three types of relationships in a relational database:

- One-to-many   The most common type of relationship is one-to-many. This means that for each occurrence in a given entity, the parent entity, there may be one or more occurrences in a second entity, the child entity, to which it is related. For example, in the widget company database, the DEPT entity is a parent entity, and for each department, there could be one or more employees associated with that department. The relationship between DEPT and EMP is one-to-many.

- One-to-one   In a one-to-one relationship, a row in a table is related to only one or none of the rows in a second table. This relationship type is often used for subtyping. For example, an EMPLOYEE table may hold the information common to all employees, while the FULLTIME, PARTTIME, and CONTRACTOR tables hold information unique to full-time employees, part-time employees, and contractors, respectively. These entities would be considered subtypes of an EMPLOYEE and maintain a one-to-one relationship with the EMPLOYEE table. These relationships are not as common as one-to-many relationships, because if one entity has an occurrence for a corresponding row in another entity, in most cases, the attributes from both entities should be in a single entity.

- Many-to-many   [2]In a many-to-many relationship, one row of a table may be related to many rows of another table, and vice versa. Usually, when this relationship is implemented in the database, a third entity is defined as an intersection table to contain the associations between the two entities in the relationship. For example, in a database used for school class enrollment, the STUDENT table has a many-to-many relationship with the CLASS table—one student may take one or more classes, and a given class may have one or more students. The intersection table STUDENT_CLASS would contain the combinations of STUDENT and CLASS to track which students are in which classes.

Assigning Attributes to Entities

Once the designer has defined the entity relationships, the next step is to assign the attributes to each entity. This is physically implemented using columns, as shown here for the SALGRADE table as derived from the salary grade entity.

GRADE	LOSAL	HISAL
1	700	1200
2	1201	1400
3	1401	2000
4	2001	3000
5	3001	9999
6	10000	12500

Iterate the Process: Are We There Yet?

After the entities, relationships, and attributes have been defined, the designer may iterate the data modeling many more times. When reviewing relationships, new entities may be discovered. For example, when discussing the widget inventory table and its relationship to a customer order, the need for a shipping restrictions table may arise.

Once the design process is complete, the physical database tables may be created. Logical database design sessions should not involve physical implementation issues, but once the design has gone through an iteration or two, it's the DBA's job to bring the designers "down to earth". As a result, the design may need to be revisited to balance the ideal database implementation versus the realities of budgets and schedules.

Words

attribute	n.	属性
bonus	n.	奖金
conceptualize	v.	使有概念，构思
derive	v.	得自，起源
enrollment	n.	登记，注册，入伍，入会，入学
iterate	v.	反复说，重申，重述
restrict	v.	约束，限制

Phrases

| correspond to | | 相应，符合 |
| vice versa | | 反之亦然 |

Notes

[1] 例句: In the widget company database, there is most likely a relationship between EMP and DEPT, but depending on the business rules, it is unlikely that the DEPT and SALGRADE entities are related.

分析：本句为并列句。在后一个分句中 it is unlikely that…意思是"……是不太可能的"，it 是形式主语，后面的 that 从句是真正主语。

译文：在饰品公司的数据库中，EMP 与 DEPT 之间很可能有关联，但根据业务规则，DEPT 和 SALGRADE 实体之间不太可能有关联。

[2] 例句：In a many-to-many relationship, one row of a table may be related to many rows of another table, and vice versa.

分析：本句为并列句，后一个分句中的 vice versa 意思是"反过来也是一样，反之亦然"。

译文：在多对多关联中，一个表的一行可能对应另一个表的许多行，反之亦然。

Exercises

Ⅰ. Put "true" or "false" in the brackets for the following statements according to the passage.

1. (　　) In data modeling process, the developer conceptualizes and documents all the tables for the database.

2. (　　) One of the common methods for modeling a database is called ERA, which stands for entities, rows, and attributes.

3. (　　) In general, an entity corresponds to a table in the database, and the attributes of the entity correspond to rows of the table.

4. (　　) An associative table resolves a many-to-many relationship.

5. (　　) In general, there are six types of relationship in a relational database.

6. (　　) Once the designer has defined the entity relationships, the next step is to assign the attributes to each entity.

7. (　　) The data-modeling process involves defining the entities, defining the relationships between those entities, and then defining the attributes for each of the entities.

8. (　　) In general, there are two types of relationship in a relational database.

9. (　　) Once the designer has defined the entity relationships, the next step is to assign the name to each entity.

10. (　　) Once the design process is complete, the logical database tables may be created.

Ⅱ. Fill in the blanks according to the passage.

1. One of the common methods for modeling a database is called ERA, which stands for _____, _____, and _____.

2. In general, an entity corresponds to a table in the database, and the _____ of the entity correspond to columns of the table.

3. The most common type of relationship is _____.

4. In a _____ relationship, a row in a table is related to only one or none of the rows in a second table.

5. In a many-to-many relationship, one row of a table may be related to _____ of another table

6. _____ is a database table that stores the valid combinations of rows from two

other tables and usually enforces a business rule. An associative table resolves a many-to-many relationship.

7. In general, there are three types of relationships in a relational database: _____, _____, and _____.

8. After the _____, _____ and _____ have been defined, the designer may iterate the data modeling many more times.

9. _____ is a process of defining the entities, attributes, and relationships between the entities in preparation for creating the physical database.

10. Logical database design sessions should not involve _____, but once the design has gone through an iteration or two, it's the DBA's job to bring the designers "down to earth".

III. Translate the following words and expressions into Chinese

1. associative table
2. attributes to entities
3. entity relationship
4. Data Modeling
5. occurrence
6. intersection table
7. derive from
8. restriction
9. versus
10. budgets and schedules

4.3.1 Reading Material

Summary of Database System

A database management system includes the physical files in which the data are stored, the software that supports access to and modification of the data, and the database schema that specifies the logical layout of the database. The relational model is the most popular database approach today. It is based on organizing data into tables of records (or objects) with particular fields (or attributes). A key field(s), whose values uniquely identify individual records in the table, is usually designated for each table.

Relationships among database elements are represented in new tables that may have their own attributes. Relationship tables do not duplicate data in other tables. Instead they store the key values of the appropriate database records so that the detailed data can be looked up when needed.

The Structured Query Language (SQL) is the standard database language for querying and manipulating relational databases. The select statement is used for queries and has many variations so that particular data can be accessed from the database. Other SQL statements allow data to be added, updated, and deleted from a database.

A database should be carefully designed. Entity-relationship modeling, with its associated ER diagrams, is a popular technique for database design. ER diagrams graphically depict the relationships among database objects and show their attributes and cardinality constraints.

Words

designate   v.   指定(出示)，标明，选派，任命 (to, for)

particular	adj.	特殊的，特别的，独特的，详细的，精确的
represent	v.	表现，描绘，声称，象征
uniquely	adv.	独特地，唯一地，珍奇地
variation	n.	变更，变化，变异，变体

Abbreviations

SQL (Structured Query Language)　　　　结构化查询语言

4.3.2　正文参考译文

关系数据库的结构（二）

数据建模

斯科特在创建数据库的真实表之前，要经过一个被称作数据建模的过程。在这个过程中，数据库创建者定义和填写数据库中的所有表。ERA 是为数据库建模的一种常用方式，ERA 可以表示出实体、实体间的关联和实体的属性。数据库设计者使用一个能够支持实体、实体属性和实体间关联的应用程序。通常，一个实体对应数据库中的一个表，而实体的属性对应于表中的列。

数据建模：定义实体、实体属性和实体间关联，为建立物理数据库做准备。

数据建模过程包括定义实体、定义实体间关联以及定义每个实体的属性。一个周期完成后需要不断重复，直到设计者抓住了足以建立数据库的重点。让我们进一步了解数据建模过程的各个步骤。

定义实体

首先，设计者确定数据库应用程序范围内的所有实体。实体是机构中重要的且需要记录在数据库中的人、地方或事物。实体将巧妙地转化为数据表。例如，在斯科特饰品公司数据库第一版中，斯科特定义了四个实体：员工、部门、工资水平和奖金。这些实体将成为 EMP(员工)表、DEPT(部门)表、SALGRADE(工资水平)表和 BONUS(奖金)表。

定义实体间的关联

一旦定义了实体，设计者就能够继续定义每个实体间是如何关联的。通常，设计者将每个实体同其他实体配对，并且考虑："两者之间是否存在关联？"实体间的某些关联是显而易见的，某些不是。

在饰品公司的数据库中，EMP 与 DEPT 之间很可能有关联，但根据业务规则，DEPT 和 SALGRADE 实体之间不太可能有关联。如果业务规则是用来约束某个部门的工资水平的，就可能需要一个新的实体来说明工资水平和部门之间的关联。这个实体被称作关联表或相交表，其中包含工资水平和部门之间的有效联合。

关联表：是一个数据库表，其中保存着另外两个表的行(记录)间的有效结合，通常强制执行业务规则。关联表处理多对多关联。

通常，关系数据库中有三种关联方式：

- 一对多关联 最常见的关联是一对多关联。一对多关联意思是：对于每个给出的现有实体(即父实体)都有一个或多个现有的另一个实体(即子实体)与之相关联。例如，在饰品公司数据库中，部门实体是一个父实体；而对于每个部门，都有一个或多个员工与该部门有关系。这样，部门实体和员工实体间的关联就是一对多关联。
- 一对一关联 在一个一对一关联中，表中的一行只关联另一个表中的一行或者 0 行。这种关联类型通常用于子类型。例如，一个员工表可能保存了所有员工的信息，而**全职表**、**兼职表**和**承包人表**则分别只保存全职员工、兼职员工和承包人的信息。这些实体被认为是员工表的子表，并且同员工表维持一对一关联。这种关联不像一对多关联那么常见，因为如果一个实体与另一个实体总有对应行，在大多数情况下，两个实体中的属性只在一个实体内出现就可以了。
- 多对多关联 在多对多关联中，一个表的一行可能对应另一个表的许多行，反之亦然。通常，在数据库中执行这些关联时，往往再定义第三个实体作为关联表用来保存前两个实体间的所有关联。例如，在一个课程注册数据库中，学生表与课程表之间有一个多对多关联——一个学生可能听一门或多门课程，并且一个课程也可能有一个或多个学生选修。而学生——班级关系表中就包含了学生和课程之间的关系，以表明哪个学生上哪个课。

指定实体属性

一旦设计者定义了实体间关联，下一步就是去指定每个实体的属性。这使用列来实现，右图所示为根据工资水平实体所建立的工资水平表。

等级	最低标准	最高标准
1	700	1200
2	1201	1400
3	1401	2000
4	2001	3000
5	3001	9999
6	10000	12500

重复步骤：我们达到目标了么？

在定义了实体、关联以及属性之后，设计者往往要多次重复数据建模过程。回顾关联时，可能会发现新的实体。比如，当讨论饰品库存表以及它与客户订单的关联时，就会发现需要制定一个送货约束表。

一旦设计过程完成，将要建立实际的数据库表。逻辑数据库的设计过程不会牵涉实际执行中的问题。但是当设计进入到实际的反复运作，数据库管理员就会让设计者从理想回到现实中来。结果，就可能需要再次构想设计，以求得理想的数据库运行与现实的预算和进度之间的平衡。

4.3.3 阅读材料参考译文

数据库系统小结

一个数据库管理系统包括存储数据的物理文件、支持对数据进行存取和修改的软件、以及描述数据库逻辑布局的数据库方案。关系模型是当今最常用的数据库实现方法，它是基于将数据组织成有特殊字段(或属性)的记录(或对象)的表格。关键字段的值唯一地标识表格中的一个记录，因此要为每个表格指明关键字段。

数据库要素之间的关系表示成新的表格，可能有自己的属性。关系表不复制其他表中的数据，而是存储适当的数据库记录的关键字段值，以便在需要时查阅详细资料。

结构化查询语言(SQL)是查询和操作关系数据库的标准数据库语言。Select 语句用于查询并且具有很多变化形式，使得可以从数据库中存取特殊数据。其他 SQL 语句允许对数据库中的数据进行添加、更新、删除。

我们应该精心设计数据库。实体关系建模具有相关的 ER 图，是数据库设计的热门技术。ER 图生动地描绘数据库对象之间的关系，并表示它们的属性和基数性约束。

4.4 Structured Query Language

Consider the database table shown in Figure 4.2, which contains information about movies. Each row in the table corresponds to a record. Each record in the table is made up of the same fields in which particular values are stored. That is, each movie record has a MovieId, a Title, a Genre and a Rating that contain the specific data for each record. A database table is given a name, such as Movie in this case.

MovieId	Title	Genre	Rating
101	Sixth Sense,The	Thriller horror	PG-13
102	Back to the Future	Comedy adventure	PG
103	Monsters,Inc	Animation comedy	G
104	Alien	Sci-fi horror	R

Figure 4.2　A database table, made up of records and fields

Suppose we wanted to create a movie rental business. In addition to the list of movies for rent, we must create a database table to hold information about our customers. The Customer table in Figure 4.3 could represent this information.

CustomerId	Name	Address	CreditCardNumber
101	Dennis Cook	123 Main Street	1111 1111 1111 1111
102	Doug Nickle	456 Second Ave	2222 2222 2222 2222
103	Randy Wolf	789 Elm Street	3333 3333 3333 3333

Figure 4.3　A database table containing customer data

We can use SQL to retrieve selective data, update and change data, delete data etc. from these tables.

The Structured Query Language (SQL) is a comprehensive database language for managing relational databases. It includes statements that specify database schemas as well as statements that add, modify, and delete database content. [1]It also includes, as its name implies, the ability to query the database to retrieve specific data.

The original version of SQL was Sequal, developed by IBM in the early 1970s. In 1986, the American National Standards Institute (ANSI) published the SQL standard, the basis for commercial database languages for accessing relational databases.

SQL is not case sensitive, so keywords, table names, and attribute names can be uppercase, lowercase, or mixed. Spaces are used as separators in a statement.

Queries

Let's first focus on simple queries. The select statement is the primary tool for this purpose. The basic select statement includes a select clause, a from clause, and a where clause:

select attribute-list from table-list where condition

The select clause determines what attributes are returned. The from clause determines what tables are used in the query.   The where clause restricts the data that is returned. For example:

select Title from Movie where Rating = 'PG'

The result of this query is a list of all titles from the Movie table that have a rating of PG. The where clause can be eliminated if no special restrictions are necessary:

select Name, Address from Customer

This query returns the name and address of all customers in the Customer table. An asterisk (*) can be used in the select clause to denote that all attributes in the selected records should be returned:

select * from Movie where Genre like '%action%'

This query returns all attributes of records from the Movie table in which the Genre attribute contains the word 'action'. The like operator in SQL performs some simple pattern matching on strings, and the % symbol matches any string.

Select statements can also dictate how the results of the query should be sorted using the order by clause:

select * from Movie where Rating ='R' order by Title

This query returns all attributes of R-rated movies sorted by the movie title.

There are many more variations on select statements supported by SQL than those we've shown here. Remember that our goal is to introduce the database concepts to you. You would require much more detail to truly become proficient at SQL queries.

Modifying Database Content

The insert, update, and delete statements in SQL allow the data in a table to be changed. The insert statement adds a new record to a table. Each insert statement specifies the values of the attributes for the new record. For example:

insert into Customer values (9876, 'John Smith', '602 Greenbriar Court', '2938 3212 3402 0299')

This statement inserts a new record into the Customer table with the specified attributes.

The update statement changes the values in one or more records of a table. For example:

update Movie set Genre = 'thriller drama' where title = 'Unbreakable'

This statement changes the Genre of the movie Unbreakable to 'thriller drama'.

The delete statement removes all records from a table matching the specified condition. For example, if we wanted to remove all R-rated movies from the Movie table, we could use the following delete statement:

delete from Movie where Rating = 'R'

[2]As with the select statement, there are many variations of the insert, update, and delete statements as well.

Words

asterisk	n.	星号
clause	n.	子句，条款
commercial	adj.	商业的，贸易的
comprehensive	adj.	全面的，广泛的，能充分理解的，包容的
denote	v.	指示，表示
drama	n.	(在舞台上演)戏剧，戏剧艺术
eliminate	v.	排除，消除，除去
genre	n.	类型，流派
pattern	n.	模范，式样，模式，样品，格调，图案
	v.	仿造，以图案装饰，形成图案
primary	adj.	第一位的，主要的，初步的，初级的，原来的
proficient	n.	精通
rating	n.	等级级别(尤指军阶)，额定
rent	v.	租，租借，出租
	n.	租金
restriction	n.	限制，约束
retrieve	v.	重新得到
selective	adj.	选择的，选择性的
separator	n.	分隔符，隔离物，分离者
specify	v.	指定，详细说明，列入清单
thriller	n.	惊险读物，电影，戏剧

Phrases

as well as	adv.	也，又
in addition to	adv.	除……之外

Abbreviations

ANSI (American National Standards Institute)　美国国家标准化组织

Chapter 4　Database Fundamentals

Notes

[1] 例句：It also includes, as its name implies, the ability to query the database to retrieve specific data.

分析：本句是复合句。句中非限制性定语从句 as its name implies 作插入语，as 指代整个主句内容。

译文：顾名思义，它还包括能够查询数据库来检索特定数据的功能。

[2] 例句：As with the select statement, there are many variations of the insert, update, and delete statements as well.

分析：本句是简单句。句中 as with 与……一样，例如：One day, as with all tyrannies, Myanmar's will fall. 终有一天，就像所有的专制政权一样，缅甸军人政权会崩溃。

译文：与 select 语句一样，insert、update 和 delete 语句也有许多不同的形式。

Exercises

Ⅰ. Put "true" or "false" in the brackets for the following statements according to the passage.

1. (　　) The Structured Query Language (SQL) is a comprehensive database language for managing relational databases.
2. (　　) The SQL includes statements that add, modify, and delete database content.
3. (　　) The original version of SQL was Sequal, developed by IBM in 1986.
4. (　　) SQL is case sensitive, so keywords, table names, and attribute names must be lowercase.
5. (　　) The basic select statement includes a select clause, a from clause, and a where clause.
6. (　　) The where clause determines what tables are used in the query.
7. (　　) The from clause can be eliminated if no special restrictions are necessary.
8. (　　) The like operator in SQL performs some simple pattern matching on strings.
9. (　　) Select statements can also dictate how the results of the query should be sorted using the order by clause.
10. (　　) The delete statement adds a new record to a table.

Ⅱ. Fill in the blanks according to the passage.

1. The Structured Query Language (SQL) is a comprehensive database language for _____.
2. In 1986, the _____ (ANSI) published the SQL standard.
3. The _____ is the primary tool for query.
4. The _____ clause determines what attributes are returned.
5. The _____ clause determines what tables are used in the query.
6. The _____ clause restricts the data that is returned.
7. Select statements can also dictate how the results of the query should be sorted using the

_____ clause.

8. The _____ statement adds a new record to a table.

9. The update statement changes the _____ in one or more records of a table.

10. The _____ statement removes all records from a table matching the specified condition.

III. Translate the following words and expressions into Chinese.

1. Structured Query Language
2. relational database
3. American National Standards Institute
4. case sensitive
5. clause
6. variation
7. update
8. specify
9. restrict
10. query

4.4.1 Reading Material

PostgreSQL

PostgreSQL is an Object-Relational Database Management System (ORDBMS) based on POSTGRES, Version 4.2, developed at the University of California at Berkeley Computer Science Department. POSTGRES pioneered many concepts that only became available in some commercial database systems much later.

Features PostgreSQL is an open-source descendant of this original Berkeley code. It supports SQL92 and SQL99 and offers many modern features:

- Complex Queries
- Foreign Keys
- Triggers
- Views
- Transactional Integrity
- Multiversion Concurrency Control

Additionally, PostgreSQL can be extended by the user in many ways, for example, by adding new:

- Data Types
- Functions
- Operators
- Aggregate Functions
- Index Methods
- Procedural Languages

And because of the liberal license, PostgreSQL can be used, modified, and distributed by everyone free of charge for any purpose, be it private, commercial, or academic.

Advantages PostgreSQL offers many advantages for your company or business over other database systems.

1) Immunity to over-deployment

Over-deployment is what some proprietary database vendors regard as their #1 licence

Chapter 4 Database Fundamentals

compliance problem. With PostgreSQL, no-one can sue you for breaking licensing agreements, as there is no associated licensing cost for the software.

This has several additional advantages:
- More profitable business models with wide-scale deployment.
- No possibility of being audited for license compliance at any stage.
- Flexibility to do concept research and trial deployments without needing to include additional licensing costs.

2) Better support than the proprietary vendors

In addition to our strong support offerings, we have a vibrant community of PostgreSQL professionals and enthusiasts that your staff can draw upon and contribute to.

3) Significant saving on staffing costs

Our software has been designed and created to have much lower maintenance and tuning requirements than the leading proprietary databases, yet still retain all of the features, stability, and performance.

In addition to this our training programs are generally regarded as being far more cost effective, manageable, and practical in the real world than that of the leading proprietary database vendors.

4) Legendary reliability and stability

Unlike many proprietary databases, it is extremely common for companies to report that PostgreSQL has never, ever crashed for them in several years of high activity operation. Not even once. It just works.

5) Extensible

The source code is available to all at no charge. If your staff have a need to customise or extend PostgreSQL in any way then they are able to do so with a minimum of effort, and with no attached costs. This is complemented by the community of PostgreSQL professionals and enthusiasts around the globe that also actively extend PostgreSQL on a daily basis.

6) Cross platform

PostgreSQL is available for almost every brand of Unix platforms with the latest stable release), and Windows compatibility is available via the Cygwin framework. Native Windows compatibility is also available with version 8.0 and above.

7) Designed for high volume environments

We use a multiple row data storage strategy called MVCC to make PostgreSQL extremely responsive in high volume environments. The leading proprietary database vendor uses this technology as well, for the same reasons.

8) GUI database design and administration tools

Several high quality GUI tools exist to both administer the database (pgAdmin, pgAccess) and do database design (Tora, Data Architect).

Words

aggregate	v.	集合，聚集
Berkeley	n.	伯克利(美国加利福尼亚州西部城市)
descendant	n.	子孙，后裔，后代
enthusiast	n.	热心家，狂热者
extensible	adj.	可扩张的，可扩展的
framework	n.	构架，框架，结构
integrity	n.	完整性
immunity	n.	免疫性
legendary	adj.	传说中的
trigger	v.	引起，触发

Abbreviations

GUI(Graphical User Interface)　　　　图形用户界面
SQL(Structured Query Language)　　　结构化查询语言
ORDBMS(Object-Relational Database 　对象–关系型数据库管理系统
　　　　Management System)

4.4.2　正文参考译文

结构化查询语言

看看图 4.2 所示的数据库表，该表包含有关影片的信息。表中的每一行对应一个记录，表中的每个记录由相同的存储特定值的字段组成。也就是说，每个影片记录有影片标识、名称、流派、分级，这些字段包含每个记录的特定数据。每个数据库表应该有一个名字，例如该表的名字为 Movie。

假设我们想建立一个影片租赁业务。除了要有出租的影片列表之外，我们还必须建立一个数据库表以保存我们的客户信息。图 4.3 中的 Customer 表可以表示这方面的资料。

我们可以使用 SQL 对这些表格进行选择性数据检索、更新和改变数据、删除数据等。

结构化查询语言(SQL)是一个用于管理关系数据库的全面的数据库语言。它包括描述数据库方案的语句以及添加、修改和删除数据库内容的语句。顾名思义，它还包括能够通过查询数据库来检索特定数据的功能。

SQL 的最初版本是 Sequal，20 世纪 70 年代初由 IBM 开发。1986 年，美国国家标准学会(ANSI)公布 SQL 标准，SQL 标准是访问关系数据库的商业数据库语言的基础。

SQL 是不区分大小写的，因此，关键字、表名和属性名称可以大写、小写或大小写混合。空格被用作为语句中的分隔符。

查询

让我们首先看看简单查询。select 语句是实现这一目的的主要手段。基本的 select 语句包含一个 select 子句、一个 from 子句和一个 where 子句：

select 属性列表 from 表名列表 where 条件

select 子句确定返回的属性，from 子句确定查询用到哪些表，where 子句限制返回的数据。例如：

select Title from Movie where Rating = 'PG'

这个查询的结果是 Movie 数据表中的分级是 PG 的所有名称的列表。如果没有必需的特殊限制，where 子句可以省略：

select Name, Address from Customer

此查询返回客户表中所有客户的名称和地址。星号(*)可用于 select 子句中表示需要返回选中记录的所有属性：

select * from Movie where Genre like '%action%'

此查询将返回 Movie 表中的 Genre 属性包含单词 action 的记录的所有属性。SQL 的 like 操作符对字符串执行一些简单的模式匹配，%符号匹配任何字符串。使用 order by 子句，select 语句可以决定查询的结果应该如何排序：

select * from Movie where Rating ='R' order by Title

此查询将返回按影片名排序的 R 级电影的所有属性。

相对于已经列出来的，SQL 支持的 select 语句还有更多的变化形式。请记住，本节的目标是介绍数据库的概念。为了真正精通 SQL 查询，将需要了解更多的细节。

修改数据库内容

SQL 的 insert、update 和 delete 语句允许改变数据表中的数据。insert 语句向表中增加一条新的记录。每个 insert 语句指定新记录的属性的值。例如：

insert into Customer values (9876, 'John Smith', '602 Greenbriar Court', '2938 3212 3402 0299')

该语句插入一个具有指定属性的新记录到 Customer 表。

update 语句改变表的一个或多个记录的值。例如：

update Movie set Genre = 'thriller drama' where title = 'Unbreakable'

该语句将 Movie 表中的 Title(名称)为' Unbreakable'(《坚不可摧》)的 Genre(流派)属性改为'thriller drama'(惊险剧)。

Delete 语句删除数据库中所有符合特定条件的记录。例如，如果想从 Movie 表中删除所有 R 级影片，可以使用下面 delete 语句：

delete from Movie where Rating = 'R'

与 select 语句一样，insert、update 和 delete 语句也有许多不同形式。

4.4.3 阅读材料参考译文

PostgreSQL

PostgreSQL 是一种基于 POSTGRES 的对象-关系数据库管理系统(ORDBMS)，

PostgreSQL 4.2 版是加利福尼亚大学的伯克利计算机系研发的。POSTGRES 开创出许多很久以后才在某些商业数据库中有效的概念。

特征 PostgreSQL 是最初伯克利代码中的开源分支，支持 SQL92 和 SQL99，并且提供了许多新功能：
- 复杂查询
- 外键
- 触发
- 查看
- 事务完整性
- 多版本并发控制

另外，用户能够通过许多方式扩展 PostgreSQL，比如通过添加新的：
- 数据类型
- 函数
- 算子
- 聚合函数
- 索引方式
- 过程语言

由于其开放的许可证，PostgreSQL 能够被所有人以任何私人、商业或学术的目的免费使用、修改和分配。

优势 对公司或行业而言，PostgreSQL 与其他数据库系统相比有许多优势。

1) 对过度部署的免疫能力

过度部署是指某些专有数据库供应商把什么看作他们的 #1 许可证遵守的问题。用 PostgreSQL，没有人能因为破坏许可协议而起诉你，因为该软件没有相关的授权费用。

这有另外几个好处：
- 可以大规模部署更多的商业盈利模式。
- 在任何阶段不可能因为许可证遵守问题被审计。
- 可以灵活地进行概念研究及测试部署而不需要额外的许可费用。

2) 比专有供应商更好的支持

除了强大的支持服务，我们还有一个由 PostgreSQL 的专业人士和爱好者组成的充满活力的团队，您的员工可以借鉴我们团队的成果，为团队做出贡献。

3) 大笔节约职员开销

比起主流专用数据库，我们所设计开发的软件有着更低的维护费和调整需求，而它仍然包含了所有的特征、稳定性和操控性。

另外，与领先的专有数据库厂商相比，我们的培训课程通常被认为更有效益、更易于管理、在现实中更实用。

4) 出名的可靠性和稳定性

与许多专有数据库不同，使用 PostgreSQL 的公司常常报告称在数年的高活性操作中，PostgreSQL 从来没有突然崩溃过。

5) 可扩展性强

它的源代码是完全免费和开放的。如果你的职员需要以任何方式定制或扩展 PostgreSQL，他们可以花最小的力气做到，并且没有额外花销。它的源代码得到全球 PostgreSQL 专家和爱好者组成的团队的补充，该团队每天都在积极扩展 PostgreSQL。

6) 跨平台

PostgreSQL 几乎可以使用在任何品牌的具有最新可靠版本的 Unix 平台上，并且通过 Cygwin 框架具备了 Windows 兼容性。它也可以兼容 8.0 及以上版本的本地 Windows。

7) 针对高性能环境的设计

我们使用称作 MVCC 的多行数据存储模式来使 PostgreSQL 能够在高性能环境中快速响应。出于同样的原因，主流专有数据库供应商也使用这种技术。

8) GUI 数据库设计和管理工具

许多高性能的 GUI 工具既可以用来进行数据库管理(如 pgAdmin, pgAccess)，也可以用来进行数据库设计(如 Tora, Data Architect)。

Chapter 5　Programming Language

5.1　Algorithms and Flowcharts

The computer scientist Niklaus Wirth stated that:

Programs = Algorithms + Data

The algorithm is part of the blueprint or plan for the computer program; an algorithm is: "An effective procedure for solving a problem in a finite number of steps."

It is effective, which means that an answer is found and it finishes, that is it has a finite number of steps. A well-designed algorithm will always provide an answer, and it may not be the answer you want but there will be an answer. It may be that the answer is that there is no answer. A well-designed algorithm is also guaranteed to terminate.

The key features of an algorithm are:

Sequence (also known as Process)

Decision (also known as Selection)

Repetition (also known as Iteration or Looping)

In 1964 the mathematicians Corrado Bohm and Guiseppe Jacopini demonstrated that any algorithm can be stated using sequence, decision and repetition. The work of Bohm and Jacopini was of great importance since it eventually led to the disciplines of structured program design that are much used today.

Sequence means that each step or process in the algorithm is executed in the specified order. In an algorithm each process must be in the correct place, otherwise the algorithm will most probably fail.

The Decision constructs—If ... then ... , If ... then ... else ...

In algorithms the outcome of a decision is either true or false, and there is no in between. [1]The outcome of the decision is based on some condition that can only result in a true or false value.

The decision takes the form: if proposition then process

A proposition in this sense is a statement, which can only be true or false. It is either true that today is Wednesday or false that today is Wednesday. It can't be both true and false. If the proposition is true then the process, which follows the then, is executed.

The decision can also be stated as:

if proposition
then process1
else process2

Chapter 5 Programming Language

This is the if ... then ... else ... form of the decision. This means that if the proposition is true then execute process1 else or otherwise execute process2.

The first form of the decision if proposition then process has a null else, that is, there is no else.

The Repetition constructs—Repeat and While

Repetition takes two forms, the Repeat loop and the While loop.

The repeat loop is used to iterate or repeat a process or sequence of processes until some condition becomes true. It has the general form:

Repeat

Process1

Process2

......

ProcessN

Until proposition

The repeat loop does some processing before testing the state of the proposition.

The while loop is used to iterate or repeat a process or sequence of processes while some condition becomes true. It has the general form:

While   proposition

Process1

Process2

......

ProcessN

The while loop tests the state of the proposition first.

There are four different ways of stating algorithms: Step-Form, Pseudocode, Flowchart, and Nassi-Schneiderman.

The first two are written forms. The written form is just a normal language. [2]A problem with human language is that it can seem to be imprecise. In terms of meaning, what I write may not be the same as what you read. [3]Pseudocode is also human language but tends toward more precision by using a limited vocabulary.

The last two are graphically-oriented, that is they use symbols and language to represent sequence, decision and repetition.

Flow charts are a graphical method of designing programs and once the rules are learned they are very easy to draw. A well-drawn flow chart is also very easy to read. Figure 5.1 shows the Basic Flowchart Symbols.

Figure 5.1   Basic Flowchart Symbols

The major symbols are the DECISION (also known as selection) and the SEQUENCE (or process) symbols. The START and STOP symbols are called the terminals. The SUBPROCESS symbol is a variation on the sequence symbol.

Words

algorithm	n.	[数]算法
decision	n.	决定，判断
discipline	n.	纪律，学科
	v.	训练
iteration	n.	迭代，重复，反复
procedure	n.	程序
pseudocode	n.	伪代码
repetition	n.	重复，循环
sequence	n.	顺序，序列
structure	n.	结构
	v.	构成，组织
terminate	v.	停止，结束

Notes

[1] 例句：The outcome of the decision is based on some condition that can only result in a true or false value.

分析：本句为复合句。定语从句中的 result in..., 意为"产生某种作用或结果"。例如：Our effort resulted in success. 我们的努力终于成功了。

译文：判断的结果基于某些条件，条件的结果只可能是真值或假值。

[2] 例句：A problem with human language is that it can seem to be imprecise.

分析：本句是复合句。句中 that it can seem to be imprecise 为表语从句，作 is 的表语，其中 it 指代"人类语言"。

译文：使用人类语言的问题是它可能看起来不精确。

[3] 例句：Pseudocode is also human language but tends toward more precision by using a limited vocabulary.

分析：本句是简单句。句中 tends towards..., 意为"倾向……"，"趋于……"；by using a limited vocabulary 是介词短语作状语，修饰 tends toward more precision。

译文：虽然伪代码也是人类语言，但它通过使用特定的词汇而倾向于更精确。

Exercises

Ⅰ. Put "true" or "false" in the brackets for the following statements according to the passage.

1. (　) Any algorithm can be stated using sequence, decision and repetition.

2. (　) Sequence means that each step or process in the algorithm is executed in the specified order.

3. (　) In algorithms, the outcome of a decision is Neither true Nor false.

Chapter 5 Programming Language

4. ( ) If the proposition is false, then the process which follows the then is executed.

5. ( ) The repeat loop does some processing after testing the state of the proposition.

6. ( ) The while loop does some processing after testing the state of the proposition or does not do anything.

7. ( ) Decision is also known as Iteration.

8. ( ) Repetition is also known as Selection or Looping.

9. ( ) The outcome of the decision is based on some condition that can only result in a true or false value.

10. ( ) An algorithm may be written down as a flowchart.

II. Fill in the blanks according to the passage.

1. _____ is: An effective procedure for solving a problem in a finite number of steps.

2. An algorithm may be written down as _____ for a computer, or as a _____.

3. A well-designed algorithm is guaranteed to _____.

4. Sequence is the key feature of an _____.

5. Decision boxes are _____ shaped.

6. Flowchart is _____-oriented.

7. Human language can seem to be _____.

8. Pseudocode is also human language but tends toward more _____ by using a limited vocabulary.

9. A _____ flow chart is very easy to read.

10. In algorithms the outcome of a _____ is either true or false.

III. Translate the following words and expressions into Chinese.

1. scientist
2. program
3. finite
4. condition
5. importance
6. execute
7. order
8. correct
9. construct
10. well-designed

5.1.1 Reading Material

Top-Down Design

The top-down design process starts by breaking the problem into a set of subproblems. Then, each subproblem is divided into subproblems. This process continues until each subproblem is defined at a level basic enough so that further decomposition is not necessary. We are creating a hierarchical structure, also known as a tree structure, of problems and subproblems, called modules. Modules at one level can call on the services of modules at a lower level. These modules are the basic building blocks of our algorithm.

The goal of dividing our problem into subproblems, modules, or segments is to be able to solve each module fairly independently of the others. In a computing context, one module could read data values, another could sum the values, another could print the sum, while still another compares the sum to the previous week's totals.

The design tree contains successive levels of refinement (see Figure 5.2). The top, or level 0, is our functional description of the problem; the lower levels are our successive refinements. So how do we divide the problem into modules?

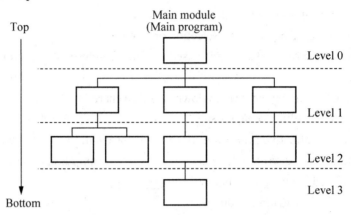

Figure 5.2   An Example of top-down design

Well, let's think for a moment about how humans usually approach any big problem. We spend some time thinking about the problem in an overall sort of way, then we jot down the major steps. We then examine each of the major steps, filling in the details. If we don't know how to accomplish a specific task, we go to the next one, planning to come back and take care of the one we skipped later when we have more information. What are we doing? We are dividing the problem into subproblems; we are using the divide and conquer strategy.

This is exactly the process you should be using in designing an algorithm. Write down the major steps. This then becomes your main module. Begin to develop the details of the major steps as level 1 modules. If you don't know how to do something, or feel overwhelmed by details, give the task a name and go on. The name can be expanded later as a lower module.

This process continues for as many levels as it takes to expand every task to the smallest details. A step that needs to be expanded is an abstract step. A step that does not need to be expanded is a concrete step. If a task is cumbersome or difficult, defer its details to a lower level. This process can be applied to the troublesome subtasks. Eventually, the whole problem is broken up into manageable units.

Writing a top-down design is similar to writing an outline for an English paper. The domain of computing is new, but the process is one you have done all your life.

Top-Down Design: A technique for developing a program in which the program is divided into more easily handled subproblems, the solutions of which create a solution to the overall problem.

Module: A self contained collection of steps that solves a problem or subproblem.

Abstract step: An algorithmic step for which some details remain unspecified.

Concrete step: A step for which the details are fully specified.

Words

cumbersome	*adj.*	讨厌的，麻烦的，笨重的
jot	*n.*	少量，少额，稍许
	vt.	略记，摘要记载下来
module	*n.*	模块
overwhelm	*vt.*	淹没，覆没，受打击，制服，压倒
subproblem	*n.*	子问题，次要问题
successive	*adj.*	继承的，连续的

5.1.2 正文参考译文

算法和流程图

计算机科学家 Niklaus Wirth 提出了这样的公式：程序=算法+数据。

算法是计算机程序的蓝图或计划的一部分，算法就是："用有限的步骤解决问题的有效的过程。"

所谓有效，就是说找到了答案并且停止了，也就是说有有限的步骤。一个设计良好的算法总是能得到答案，也许不是希望的答案但是会有答案。也可能答案本身就是没有答案。一个设计良好的算法应该保证能够终止。

算法的主要特征如下：

顺序(也称处理)、判断(也称选择)、循环(也称重复)。

数学家 Bohm 和 Jacopini 于 1964 年证明：任何一个程序均可用"顺序"、"判断"和"循环"来陈述。Bohm 和 Jacopini 的工作非常重要，因为它最终形成了当前应用广泛的结构化定理。

顺序算法就是处理过程的每一步以特定的顺序执行，在这种算法中的每一个过程要处在正确的位置，否则算法可能会失败。

判断结构——If…then…、If…then…else…

判断算法的结果或者是真或者是假，不可能有中间值。判断的结果基于某些条件，条件的结果只可能是真值或假值。

判断语句的形式为：if 条件 then 过程

此形式中的条件是一个陈述，只能是真或假。譬如：今天是星期三这个说法或者是真或者是假，不可能又是真又是假。如果条件是真，then 后的过程被执行。

判断语句还可以写成如下形式：

If 条件
then 过程 1
else 过程 2

这是 if … then … else …形式的判断语句，表示如果条件的结果为真，则执行过程 1，否则执行过程 2。

第一种形式的判断语句的 else 为空，也就是说没有 else。

循环结构——重复型循环和当型循环

循环有两种形式，重复型循环和当型循环。

重复型循环用来重复一些过程直到条件为真，一般形式如下：

(略)

这种循环在测试条件状态之前先执行过程。

当型循环当条件为真时重复一些过程，一般形式如下：

(略)

当型循环先测试条件的状态。

有四种表示算法的方法：Step-Form、伪代码、流程图和 N-S 流程图。

前两种方法是书写形式的，是普通语言。使用人类语言的问题使它可能看起来不精确，也就是说作者所写的意思和读者读到的意思可能不一致。伪代码也是人类语言，但通过使用特定的词汇而倾向于更精确地表达所述内容。

后面的两种是面向图形的，也就是说它们使用符号和语言来表示顺序、判断和循环。

流程图是程序设计的图形表示法，只要掌握了规则，就很容易画流程图，画得好的流程图也很容易读。图 5.1 所示为基本的流程图符号。

主要的符号是判断(或者选择)符号和顺序(处理)符号，开始和结束的符号称为端点符号，子过程的符号是顺序符号的变形。

5.1.3 阅读材料参考译文

自顶向下的设计

自顶向下的设计开始时要把问题分解成一组子问题，然后子问题再分解成下一级子问题。这个过程继续到每个子问题已经被在基本水平定义不需再分。建成一个问题和子问题的分级结构又名树形结构，也称为模块结构。某级别的模块可以调用下一级模块的服务。这些模块是算法的基本组成块。

把问题分解成子问题或模块或片段的目的是可以独立地解决每个模块的问题。在计算环节，可用一个模块读取数据值，另一个模块进行求和，再一个模块用于输出，而还有一个模块将得到的和与上周的总数进行比较。

树形设计包括逐次细分的层(见图 5.2)。顶层或者 0 层是对问题的功能描述；较低的层是逐次的细分。那么怎么把问题划分为模块呢？

思考一下通常我们怎样处理大问题。我们花一些时间思考整个问题，然后简单记录主要步骤。之后检查每个主要步骤，补充细节。如果不知道怎么执行某个具体任务，先去处理下一个问题，计划等到有更多信息的时候再回头处理所跳过的问题。我们在做什么？我们在把问题分解为子问题，在应用各个击破的策略。

这正好是设计算法要用的过程。写下主要步骤，这些主要步骤成为主模块。开始设计主要步骤的细节作为 1 层模块。如果不知道怎样做某些事情或者对细节不知所措，先给该任务命名然后继续。该名字后来可以展开为下一层模块。

这个过程继续进行，直到把每个任务展开为最小的细节。需要展开的步骤是抽象步骤，不需要再展开的步骤是具体步骤。如果一个任务很麻烦或很困难，把它的细节推到下一层。这个过程也可应用于复杂的子任务。最后，整个任务被分解成易处理的单元。

进行自顶向下的设计与写一篇文章的提纲类似。计算机编程是个新领域，但过程是我们曾经用过的。

自顶向下的设计：一项编程技术，把程序分解为容易处理的子问题，逐个解决子问题就解决了整个问题。

模块：解决一个问题或子问题的独立的步骤集合。

抽象步骤：一些细节未指明的算法步骤。

具体步骤：细节全部明确指出的步骤。

5.2　Introduction of Programming Languages

A programming language is a defined set of instructions that are used to make a computer perform a specific task. Written using a defined vocabulary the programming language is either complied or interpreted by the computer into the machine language that is understood by the processor.

There are several types of programming languages, the most common are:

High-level Languages—these are written using terms and vocabulary that can be understood and written in a similar manner to human language. [1] They are called high-level languages because they remove many of the complexities involved in the raw machine language that computers understand. The main advantage of high-level languages over low-level languages is that they are easier to read, write, and maintain. All high-level languages must be compiled at some stage into machine language. The first high-level programming languages were designed in the 1950s. Now there are dozens of different languages, including BASIC, COBOL, C, C++, FORTRAN and Pascal.

Scripting Languages—like high-level languages, scripting languages are written in manner similar to human language. Generally, scripting languages are easier to write than high-level languages and are interpreted rather than compiled into machine language. Scripting languages, which can be embedded within HTML, commonly are used to add functionality to a Web page, such as different menu styles or graphic displays, or to serve dynamic advertisements. These types of languages are client-side scripting languages, affecting the data that the end user sees in a browser window. Other scripting languages are server-side scripting languages that manipulate the data, usually in a database, on the server. Scripting languages came about largely because of the development of the Internet as a communications tool. Some examples of scripting languages include VBScript, JavaScript, ASP and Perl.

Assembly Language—assembly language is as close as possible to writing directly in machine language. Due to the low level nature of assembly language, it is tied directly to the type of processor and a program written for one type of CPU generally will not run on another.

Machine language—The lowest-level programming language. Machine languages are the only languages understood by computers. While easily understood by computers, machine languages are almost impossible for humans to use because they consist entirely of numbers.

Programmers, therefore, use either a high-level programming language or an assembly language. An assembly language contains the same instructions as a machine language, but the instructions and variables have names instead of being just numbers.

Programs written in high-level languages are translated into assembly language or machine language by a compiler. Assembly language programs are translated into machine language by a program called an assembler.

Every CPU has its own unique machine language. Programs must be rewritten or recompiled, therefore, to run on different types of computers.

For now, let's talk about some high level languages very briefly, which are used by professional programmers in the current mainstream software industry.

- C

This is probably the most widely-used, and definitely the oldest, of the three languages I mentioned. It's been in use since the 70s, and is a very compact (i.e. not much "vocabulary" to learn), powerful and well-understood language.

- C++

This language is a superset of C (that just means that it's C with more stuff added; it's more than C, and includes pretty much all of C). Its main benefit over C is that it's object oriented. [2]The key point is that object oriented languages are more modern, and using object oriented languages is the way things are done now in the programming world. So C++ is most definitely the second-most used language after C, and may soon become the most used language.

- Java

Java has a benefit which other programming languages lack: it's cross-platform. It means that it runs on more than one platform without needing to be recompiled. A platform is just a particular type of computer system, like Windows or Mac OS or Linux. Normally, if you wanted to use the same program on a different platform from the one it was written on, you'd have to recompile it—you'd have to compile a different version for each different platform. Sometimes, you'd also need to change some of your code to suit the new platform. This probably isn't surprising, since the different platforms work differently, and look different.

Java has another advantage that it runs inside web browsers, letting programmers create little applications which can run on web sites. However, Java also has a disadvantage, which is almost as serious: it's slow. Java achieves its cross-platform trick by putting what is essentially a big program on your computer which pretends to be a little computer all of its own. The Java runs inside this "virtual machine", which runs on whatever platform you're using (like Windows or Mac OS). Because of this extra layer between the program and the computer's processor chip, Java is slower than a program written and compiled natively for the same platform. Anyway, as the Internet develops, Java will be used more widely.

Chapter 5 Programming Language

Words

assembler	n.	汇编程序
compact	adj.	紧凑的，简洁的
compile	v.	编译
complexity	n.	复杂性
dozen	n.	一打，十二个
cross-platform		跨平台
instruction	n.	指示，指令
interpret	v.	解释，口译
JavaScript	n.	Java 描述语言
Perl	n.	一种通用编程语言，可用于网络环境
pretend	v.	假装
specific	n.	细节
	adj.	特殊的，特效的
stage	n.	舞台，活动场所，发展的进程，阶段
stuff	n.	原料，素材资料
	v.	填充
trick	n.	诡计，骗局，诀窍
	v.	欺骗，哄骗

Phrases

due to	由于，应归于
dynamic advertisements	动态的广告
Mac OS	Mac 操作系统
scripting languages	脚本语言
super set	超集，父集
virtual machine	虚拟计算机

Abbreviations

ASP(Active Server Page)   动态服务器主页
VBScript(Visual Basic Script) 描述语言

Notes

[1] 例句：They are called high-level languages because they remove many of the complexities involved in the raw machine language that computers understand.

分析：本句是复合句。原因状语从句中的 involved in the raw machine languages that computers understand 为过去分词短语，作 complexities 的后置定语。

译文：它们之所以被称为高级语言，是因为它们去除了计算机所能理解的原始机器语言所包含的复杂性。

[2] 例句：The key point is that object oriented languages are more modern, and using object oriented languages is the way things are done now in the programming world.

分析：本句是 and 连接的并列句。在后一个分句中，using object oriented languages 是动名词短语作主语，位于后面的动词用单数 is；things are done now in the programming world 为定语从句，修饰 the way。修饰 way, distance, direction 等名词的定语从句，通常可不用起连接作用的词，有时也可用 that，但这时 that 在定语从句中作状语。

译文：关键的一点是面向对象的语言更现代化，使用该语言是当今程序设计领域处理事务的方式。

Exercises

Ⅰ. Put "true" or "false" in the brackets for the following statements according to the passage.

1. (　) Assemble languages are the only languages understood by computers.
2. (　) Low-level languages are easy to read, write, and maintain.
3. (　) Every CPU has its own unique machine language.
4. (　) Java achieves its cross-platform trick by putting what is essentially a big program on your computer which pretends to be a little computer all of its own.
5. (　) The Java runs inside a little computer of its own, which runs on whatever platform you're using.
6. (　) When we write a program in a high-level language, we have to compile a different version for each different platform.
7. (　) Java has a benefit which other programming languages lack: it's cross-platform.
8. (　) C++ program can run on more than one platform without needing to be recompiled.
9. (　) A platform is just a particular type of computer system, like Windows or Mac OS.
10. (　) Assembly language consists entirely of numbers.

Ⅱ. Fill in the blanks according to the passage.

1. A programming language is a defined set of _____ that are used to make a computer perform a specific task.
2. High-level Languages are written using terms and vocabulary that can be understood and written in a manner _____ to human language.
3. An example of high level language is _____.
4. _____ can be embedded within HTML.
5. _____ is the lowest-level programming language.
6. C++ is _____.
7. _____ language is machine-oriented.
8. Java is _____ means that it runs on more than one platform without needing to be recompiled.

9. Java has an advantage that it runs inside _____.

10. Java has a disadvantage, which is almost as serious: _____.

III. Translate the following words and expressions into Chinese.

1. program
2. specific
3. processor
4. manner
5. advantage
6. generally
7. menu
8. development
9. communication
10. consist

5.2.1 Reading Material

A Brief Word About Scripting

You've perhaps heard about something called scripting, or maybe you've heard of languages like JavaScript, VBScript, Tcl and others (those languages are called scripting languages). You may thus be wondering if scripting is the same as programming, and/or what the differences are, and so on. People get quite passionate about this question, so I'm just going to cover it briefly and technically. Here are some facts:

- Scripting is essentially a type of programming.
- Scripting languages have a few minor technical differences which aren't important to discuss at this point.
- Scripting languages tend to be interpreted rather than compiled, which means that you don't need to compile them—they're compiled "on the fly" (i.e. when necessary, right before they're run). This can make it faster to program in them (since you always have the source code, and don't need to take the deliberate extra step of compiling).
- The fact that scripting languages are interpreted generally makes them slower than programming languages for intensive operations (like complex calculations).
- Scripting languages are often easier to learn than programming languages, but usually aren't as powerful or flexible.
- For programming things like applications for personal computers, you'll need to use a programming language rather than a scripting language.

Scripting languages can be excellent for beginners: they tend to be easier to learn, and they insulate you from some of the technical aspects of programming (compiling, for one). However, if you're serious about programming, you won't be able to stay with a scripting language forever—you will move on to a programming language at some point. I'd say that it's good to know a scripting language or two, and even to start with a scripting language rather than a programming language. However, there's a point of view which says that, by protecting and "hand-holding" too much, scripting languages don't properly prepare you for "serious" programming, and set you up for a bit of a learning curve when you move on to a programming language.

Words

deliberate	adj.	故意的，预有准备的
essentially	adv.	本质上，本来
insulate	v.	使绝缘，隔离
intensive	adj.	强烈的，透彻的，[语法]加强语气的
interpret	v.	解释
Tcl	n.	一种早期的脚本语言

5.2.2　正文参考译文

程序设计语言介绍

程序设计语言是一个被明确定义的指令集，用于使计算机执行特定任务。程序设计语言使用专用词汇编写，被计算机编译或解释成处理器可以识别的机器语言。

常用的程序设计语言类型如下：

高级语言——使用接近人类的语言、易理解的短语、词汇方式书写。之所以被称为高级语言，是因为它们去除了计算机所能理解的原始机器语言所包含的复杂性。高级语言相对于低级语言的主要优点是，它们易于读写和维护。所有高级语言都要在某个阶段翻译成机器语言。最早的高级语言大约出现于 20 世纪 50 年代。现在有很多种高级语言，包括 BASIC、COBOL、C、C++、FORTRAN 和 Pascal 等。

脚本语言——类似于高级语言，脚本语言用与人类语言相似的方式书写。一般来说，用脚本语言编程比用高级语言更容易，并且脚本语言采用解释方式而不是编译方式转换成机器语言。脚本语言可以被嵌入 HTML，经常被用来在网页中添加功能(如不同的菜单风格或图形显示)或提供动态广告。这种类型的语言是客户端脚本语言，它影响终端用户在浏览器窗口看到的数据。另一种是服务器端的脚本语言，操作服务器端数据库中的数据。脚本语言产生的很大原因是通信工具的 Internet 的发展。脚本语言包括 VBScript、JavaScript、ASP 和 Perl。

汇编语言——汇编语言非常接近机器语言。由于汇编语言的低级特性，它和处理器的类型密切相关，在一种 CPU 类型上编写的程序一般不能在另一种类型上运行。

机器语言——最低级的编程语言。机器语言是唯一能直接被计算机识别的语言。易于被计算机理解的同时，机器语言几乎不能被人类使用，因为它们完全由数字组成。所以程序员使用高级语言或汇编语言。汇编语言的指令与机器语言对应，但是汇编语言中的指令或变量用名字表示而不是用数字表示。

用高级语言编写的程序要通过编译器翻译成汇编语言或机器语言。汇编语言的程序通过汇编程序翻译成机器语言。

每种 CPU 有自己专用的机器语言。当用于不同类型的计算机时，程序需要被重写或重新编译。

下面简要介绍几种主流软件工业中专业程序员采用的高级语言：

- C

这可能是本书提到的三种语言中应用最广最久的一种语言。从 20 世纪 70 年代开始应用，它是一种紧凑(即不需要学习太多词汇)、功能强并且易于理解的语言。

- C++

这种语言是 C 的超集(也就是说，是 C 语言的扩充，比 C 语言增加了很多，包括 C 语言的全部)。C++相对于 C 的主要优点是面向对象。关键的一点是面向对象的语言更现代化，使用该语言是当今程序设计领域处理事务的方式。所以 C++无疑是应用广泛度仅次于 C 的语言，并且很快会成为应用最为广泛的语言。

- Java

Java 有一个别的语言所没有的优点：它是跨平台的。也就是说，它不需要重新编译就可以运行于多个平台。平台就是计算机系统的类型，比如 Windows、Mac OS 或 Linux。一般来说，如果你在一个平台上编写了程序，想到另一个平台上去运行，需要重新编译，也就是说，需要为不同的平台编译不同的版本。有时候，你还需要更改代码以适应新的平台。因为不同的平台工作方式不同，并且外观也不同。

Java 还有一个优点，就是可以运行于 Web 浏览器，使程序员创建能够运行在 Web 站点的小的应用程序。但是，Java 也有缺点，并且很严重，就是它速度慢。Java 实现跨平台的诀窍是通过在计算机上放置一个实质上很大的程序，假装有自己的小计算机。Java 运行于这个虚拟机中，而虚拟机运行于任何一种平台(像 Windows 或 Mac OS)上。因为在程序和计算机处理器芯片之间又附加了这么一层，Java 的运行比在原平台上编写并编译的程序慢。无论如何，随着 Internet 的发展，Java 会得到更多的应用。

5.2.3 阅读材料参考译文

浅谈脚本

您也许听说过所谓的脚本，或者听说过这样的语言，如 JavaScript、VBScript、Tcl 和其他语言(这些语言被称为脚本语言)。您可能会因此想知道：脚本是否与编程相同，或者它们有什么差异等。人们非常热衷于这个问题，所以我想简要地从技术上解释一下这个问题。情况如下：

- 脚本本质上是一种编程。
- 脚本语言有一些小的技术性差异，这些差异并不重要，这里不再讨论。
- 脚本语言通常是被解释而不是被编译，这意味着不需要编译它们——它们可以"即时"编译(即必要时，在即将运行时)。这可以使在脚本中编程更加快速(因为你始终拥有源代码，不需要特意采取额外的编译步骤)。
- 脚本语言通常以解释方式执行的这一事实使它们在处理密集型业务(如复杂的计算)比编程语言慢。
- 脚本语言往往比编程语言更容易学，但通常没有编程语言功能强大、灵活。
- 编写电脑应用程序时，需要使用编程语言，而不是脚本语言。

脚本语言非常适合初学者：它们往往更容易学习，而且它们可以使初学者免受编程技术方面的(如编译)的影响。不过，如果您对编程很认真，就不能永远停留在脚本语言上——在一段实践后，就会进入一种编程语言。我认为了解一个或两个脚本语言、甚至以脚本语言而不是一种编程语言入门还是不错的。然而，也有一种看法认为：由于保护和"手控"太多，脚本语言不能正确地为"认真"的编程学习做准备，而且会使您在进入编程语言的学习时走一点弯路。

5.3 Object-Oriented Programming

Object-oriented programming (OOP) refers to a special type of programming that combines data structures with functions to create re-usable objects.

Otherwise, the term object-oriented is generally used to describe a system that deals primarily with different types of objects, and where you can take the actions depends on what type of object you are manipulating. For example, an object-oriented draw program might enable you to draw many types of objects, such as circles, rectangles, triangles, etc. Applying the same action to each of these objects, however, would produce different results. If the action is Make 3D, for instance, the result would be a sphere, box, and pyramid, respectively.

Many languages support object oriented programming. In OOP data and functions are grouped together in objects (encapsulation). An object is a particular instance of a class. [1] Each object can contain different data, but all objects belonging to a class have the same functions (called methods). So you could have a program with many e-mail objects, containing different messages, but they would all have the same functionality, fixed by the email class. Objects often restrict access to the data (data hiding).

Classes are a lot like types—the exact relationship between types and classes can be complicated and varies from language to language.

Via inheritance, hierarchies of objects can share and modify particular functions. You may have code in one class that describes the features all e-mails have (a sender and a date, for example) and then, in a sub-class for e-mail containing pictures, add functions that display images. [2]Often in the program you will refer to an e-mail object as if it was the parent (super-class) because it will not matter whether the e-mail contains a picture, or sound, or just text. This code will not need to be altered when you add another sub-class of e-mail objects, containing (say) electronic cash.

Sometimes you may want an action on a super-class to produce a result that depends on what sub-class it "really is". For example, you may want to display a list of email objects and want each sub-class (text, image, etc) to display in a different colour. In many languages it is possible for the super-class to have functions that sub-classes change to suit their own purposes (polymorphism, implemented by the compiler using a technique called dynamic binding). So each email sub-class may supply an alternative to the default, printing function, with its own colour.

In many OO languages it is possible to find out what class an object is (run time type information) and even what functions are connected with it (introspection / reflection). Others, like C++ have little run time information available (at least in the standard language—individual libraries of objects can support RTTI with their own conventions).

[3]There are at least three approaches to OO languages: Methods in Classes, Multi-Methods Separate from Classes, Prototypes.

Methods in Classes

Many languages follow Smalltalk in associating functions (methods) with classes. The methods form part of a class definition and the language implementation will have (this is a low-level detail hidden from the programmer) a vtable for each class which links methods to their implementations. This indirection is necessary to allow polymorphism, but introduces a performance penalty. In some languages (C++, at least), only some methods, marked as virtual by the programmer, are treated in this way.

Multi-Methods Separate from Classes

Some languages (e.g. Common Lisp / CLOS) allow functions to specialise on the class of any variable that they are passed (multi-methods). Functions cannot be associated with one class because different versions of the function may exist for many different combinations of classes.

Prototypes

Other OO languages do away with classes completely (e.g. Self). Prototype-based languages create new objects using an existing object as an example (prototype). Apart from solving some problems with dynamic object creation, this approach also encourages delegation (function calls are passed to other objects) rather than inheritance.

Words

complicate	v.	(使)变复杂
delegation	n.	代表团，授权
encapsulation	n.	包装，封装
implementation	n.	执行
individual	n.	个体
	adj.	个别的，单独的
inheritance	n.	遗传，遗产，继承
introspection	n.	内省，反省
penalty	n.	处罚
polymorphism	n.	多形性，多态现象
prototype	n.	原型
via	prep.	通过，经由

Phrases

do away with   废除

Abbreviations

CLOS(Common LISP Object System)   公共 LISP 对象系统.
OOP(Object -Oriented Programming)   面向对象的程序设计
RTTI(Run Time Type Information)   运行时类型信息

Notes

[1] 例句：Each object can contain different data, but all objects belonging to a class have the same functions (called methods).

分析：本句是并列句，but 连接表转折意义的两个分句。后一个分句中，belonging to a class 为现在分词短语作 objects 的后置定语。

译文：每一个对象可以包含不同的数据，但属于同一类的所有对象具有同样的功能(被称为方法)。

[2] 例句：Often in the program you will refer to an e-mail object as if it was the parent (super-class) because it will not matter whether the e-mail contains a picture, or sound, or just text.

分析：本句是复合句。主句中的 as if it was the parent 为 as if 引导的虚拟语气的方式状语从句，如果表示与目前事实不符，谓语用一般过去式。在原因状语从句 because it will not matter whether the e-mail contains a picture, sound or just text. 中，it 为形式主语，而主语从句 whether the e-mail contains a picture, sound or just text 为真正主语。

译文：在程序中人们经常查阅电子邮件对象，把它看作父类；因为该邮件是否包含图片、声音或仅仅一种文本无关紧要。

[3] 例句：There are at least three approaches to OO languages: Methods in Classes, Multi-Methods Separate from Classes, Prototypes.

分析：本句是简单句。句中 approach 为名词，意为方法，后常跟 to+名词意为"对待……"的方法。冒号后的 Methods in Classes, Multi-Methods Separate from Classes, Prototypes 为 approaches 的同位语。

译文：至少有三种方法处理面向对象的语言：类的方法、与类分离的多方法、原型。

Exercises

Ⅰ. Put "true" or "false" in the brackets for the following statements according to the passage.

1. (　) OOP stands for Object-oriented programming.
2. (　) An object is a particular instance of a class.
3. (　) A class is a particular instance of an object.
4. (　) Data hiding means that data and functions are grouped together in objects.

Chapter 5 Programming Language

5. (　) Object-oriented programming (OOP) refers to a special type of programming that combines data structures with functions to create re-usable objects.
6. (　) Popular modern object oriented programming languages include Java and C++.
7. (　) Machine language is an object oriented programming language.
8. (　) Functions cannot be associated with one class.
9. (　) Prototype-based languages create new classes using an existing object as an example.
10. (　) Objects often restrict access to the data (data hiding).

II. Fill in the blanks according to the passage.

1. In OOP data and _____ are grouped together in objects.
2. An object is a particular instance of a _____.
3. Each object can contain different data, but all objects belonging to a _____ have the same functions.
4. It is possible for the super-class to have functions that sub-classes _____ to suit their own purposes.
5. _____ versions of the function may exist for many different combinations of classes.
6. Objects often restrict _____ to the data (data hiding).
7. Via inheritance, hierarchies of objects can _____ and modify particular functions.
8. OOP combines data structures with functions to create re-usable _____.
9. There are at least three approaches to OO languages: Methods in _____, Multi-Methods Separate from Classes, _____.
10. Some languages (e.g. common Lisp / CLOS) allow _____ to specialise on the class of any variable that they are _____ (multi-methods).

III. Translate the following words and expressions into Chinese.

1. data structures
2. functions
3. describe
4. message
5. exact
6. relationship
7. feature
8. electronic
9. compiler
10. convention

5.3.1 Reading Material

Guidelines for writing software documentation

Software documentation falls into a number of categories, including technical documentation and end user documentation.

Technical Documentation

Technical software documentation should describe how an application functions. As such it can act as a reference manual for users such as developers, technical architects and designers concerned with the application's function.

What to document in technical documentation?

For most applications, technical documentation can include information about some or all of the following.

Files: A list of important files within the application.

Functions and/or subroutines: Details of each function or subroutine, together with their parameters and return values.

Global variables or constants: Details of what these are used for.

How the application fits together: In the case of web applications using technologies such as PHP or ASP, it may describe which include files are used by which pages. It may also describe the modules or class libraries used by the application.

3rd party objects: In the case of applications using Microsoft technologies it may describe which 3rd party COM objects have been used.

API Reference: Details of how to use the Application Programming Interface (API) if the particular application has one.

Associated entities: It may also be useful to document related items such as the database used by a typical client-server application.

End User Documentation

End user documentation is intended for the actual users of the software application. It can take many forms, from online websites, to electronic reference guides such as Windows HTML Help as well as the more traditional printed documentation. Although end user documentation is normally prepared in a Word processor or HTML editor, various tools exist to assist with the creation of the documentation. Such tools include document conversion utilities, screenshot generators and annotators and online help compilers.

Words

annotator	n.	注解者，文中指一种文档生成工具
category	n.	种类，类别，[逻]范畴
client-server		[计]客户-服务器
concern	vt.	涉及，关系到
	n.	关心，关注
constant	n.	[数、物]常数，恒量
document	n.	公文，文件，文档
	vt.	为……提供文件(证明)
documentation	n.	文件
fit	v.	适合，安装，使适应
global	adj.	球形的，全球的，全局的
library	n.	图书馆，库

parameter	n.	参数,参量
reference	n.	涉及,参考,参考书目
subroutine	n.	[计]子程序
variable	n.	[数]变数,变量
	adj.	可变的,不定的,[数]变量的

Phrases

as such	同样地,同量地
fall into	落入,陷于(混乱、错误等),分成,属于
in the case of	在……的情况
intend for	打算供……使用

Abbreviations

ASP (Active Server Page)	动态服务器主页
PHP	一种新型的 CGI 网络程序编写语言

5.3.2 正文参考译文

面向对象程序设计

面向对象的程序设计涉及一种专用类型的程序设计,该技术将数据结构与函数相结合产生可重用的对象。

另一方面,术语"面向对象"通常用来描述一种系统,这种系统主要处理不同类型的对象,而人们所采取的行为取决于他们正处理的对象的类型。例如,一种面向对象的绘图程序能使人们绘制多种类型的对象,如圆、长方形和三角形等。但是,将同样的行为应用于不同的对象将产生不同的结果。例如,如果这一行为是制作三维物体,结果将分别形成球体、长方体和棱锥。

许多语言都支持面向对象的程序设计。在面向对象的程序设计中,数据和功能组合在一起归入对象内(封装)。对象是类的特例。每个对象可以包含不同的数据,但属于同一类的所有对象具有相同的功能(也称为方法)。于是,人们可以有一个带有许多电子邮件对象的程序,这些电子邮件对象包含有不同的信息,但具有电子邮件类固有的相同功能。对象经常限制对数据的访问(隐藏数据)。

类很像类型——它们之间的关系很复杂,而且随着语言的变化而变化。

通过继承,不同级别的对象可以共享和修改特殊的功能。可以在一个类中用代码描述所有邮件具有的特点(例如:发送者和日期),而在包含图片的子类中,增加显示图像的功能。在程序中人们经常查阅电子邮件对象,把它看作父类;因为该邮件是否包含图片、声音或仅仅一种文本并不重要。也就是说,当添加另一个邮件对象的子类(譬如包含电子现金)时,不需要改变代码。

有时候人们可能希望父类上的一个行为可以根据子类的"真实"内容产生结果。例如:你可能想显示邮件对象的列表,并且希望每个子类(文本、图像等)显示出不同的颜色。很多语言都可能具有多态性,即父类具有功能使子类为适应它们自己的目的而变化(多态性,由编译器使用动态绑定技术来实现)。这样,每个邮件子类可以提供默认打印功能的选择,并用自己的颜色显示。

在很多面向对象的语言中,可以找出一个对象属于哪个类(运行时信息),甚至与其相关的函数(内省)。其他的像 C++则没有可用的运行时信息(至少在标准语言,对象的个别库以它们自己的约定支持运行时信息)。

至少有三种方法处理面向对象的语言:类的方法、与类分离的多方法、原型。

类的方法

Smalltalk 以及其后的许多语言都将方法与类关联。方法形成类定义的一部分,并且语言实现每个类由一个虚函数表将方法与执行联系起来。为了允许多态性,这种间接是必需的,但引起了性能下降。在某些语言中(至少 C++),只有程序员标记为虚拟的某些方法用这种方式来处理。

与类分离的多方法

某些语言(例如 Common Lisp/CLOS),允许函数专门研究不同变量的类以进行函数传递(多方法),多个函数不能和一个类相关联,因为类的不同结合可能存在不同的函数版本。

原型

其他面向对象的语言完全取消了类(例如 Self)。基于原型的语言将已有的对象作为范例(原型)生成新的对象。除了解决一些动态对象产生的问题之外,这种方法还鼓励授权(函数调用被传递到其他对象)而不是继承。

5.3.3　阅读材料参考译文

软件文档编写指南

软件文档可分为若干类,包括技术文档和最终用户文档。

技术文档

技术类的软件文档应说明某个应用程序如何运行。这样,它可以作为关注应用程序功能的开发者、技术规划师和设计师等用户的参考手册。

技术文档中要提供什么?

对于大多数应用软件,技术文档资料可以包括下述的部分或全部。

文件:应用软件范围内的重要文件的清单。

函数和/或子程序:每个函数或子程序的细节,连同其参数和返回值。

全局变量或常量:详细说明它们的作用。

应用软件是怎样搭配的:在使用如 PHP 或 ASP 技术的 Web 应用程序时,此部分可以描述哪些网页使用了哪些包含文件,也可以描述应用程序使用的模块或类库。

第三方对象:在使用微软技术时,可以描述使用过哪些第三方 COM 对象。

API 参考:应用程序编程接口(API)的详细用法(如果特定应用程序有一个 API)。

相关实体：记录相关的项目(如典型的客户-服务器应用程序使用的数据库)也很有用。

最终用户文档

最终用户文档是供软件的实际用户使用的。它可以采取多种形式，从在线网站到电子参考指南(如 Windows HTML 帮助)，以及更为传统的印刷文件。虽然最终用户文档通常是用文字处理器或 HTML 编辑器制作的，但也有各种工具协助建立文档。这些工具包括文件转换工具、屏幕截取生成器和 annotators 以及在线帮助编译器。

5.4 Program Debugging and Program Maintenance

If your program exits abnormally, then there is almost certainly a logical error (a bug) in your program. 99% of programming is finding and removing these bugs. Here are some tips to help you get started.

Before going on, it is necessary to reiterate the standard OLC policy on program debugging: Do *NOT* ask OLC for help debugging a program.[1] This stock answer is intended to give you some tips on how to get started in this area; however, in general, program debugging requires more time and effort than consultants are usually able to provide.

The first step is to find the exact line where the program exits. One way of doing this is with print statements scattered through your code. For example, you might do something like this in your source code:

```
myplot(int x, int y)
{ printf("Entering myplot()\n"); fflush(stdout);
    ---- lots of code here ------
    printf("Exiting myplot()\n"); fflush(stdout);
    return; }
```

The fflush() command in C ensures that the print statement is sent to your screen immediately, and you should use it if you're using printf() for debugging purposes.

[2] Once you have narrowed down the line where your bug occurs, the next step is to find out the value of your variables at that time. You will probably find that one of your variables contains very strange values. This is the time to check that you have not done the following things:

- Assigned an integer value to a pointer variable;
- Written to a subscript that is beyond the end of an array (remember that in C array subscripts go from 0 to N−1, not from 1 to N.)

Other mistakes also cause bugs. Make sure that your loops test correctly for their end conditions, for example.

Other kinds of bugs (programs not exiting, incorrect output) are debugged using similar methods. Again, find the line where the first error occurs, and then check the values of your variables. Once you fix a bug, recompile your program, run it again, and then debug it again as necessary.

Using printf() is a primitive method of debugging, but sometimes it's the only one that will work. If your program is too big for a debugger (such as Saber or Ddbx) or if you are working on a non-Athena platform, you may not have a debugger available. Usually, though, it is quicker and easier to use a debugger. Athena has several sophisticated debugging tools available. Saber is the tool of choice for C programmers. Gdb and Dbx may also come in handy, and both of them work with Fortran as well as with C. There are stock answers that introduce Saber and Dbx, and Saber even comes with a tutorial.

It is a fact of life in program design but there seems to be always one last bug or error to be corrected. We can broadly classify the errors as:

- Syntax errors—this class of error means that a mistake is made in the language used to state the algorithm.
- Logic errors—the algorithm is syntactically correct but doesn't do what is intended.
- Data range and data type errors—the algorithm is syntactically correct and logically correct but can be threatened by the wrong kind of data or by values which are out of range.

[3] The syntax errors aren't a serious issue during the program design phase since in practice, after designing and testing the design, the program will be implemented in a computer program language and it is at this point that syntax errors become a problem. Even so syntax errors are a minor problem since the process of building the program will capture the errors. The program simply won't build until all the syntax errors are removed.

The logic errors are a much more serious problem since there is no way to eliminate these other than rigorously testing the program design.

The data errors are also serious errors and in some respects are harder to deal with than logic errors.

Once launched, the program needs to be maintained. Definition for program maintenance is that updating programs from time to time keeps abreast of changes in an organization's needs or its hardware and software. Based on the maintenance tasks needed to be performed, the program administrators should determine on-going financial and staffing needs and how they will be met. Program maintenance represents a major portion of the total expenditures on application programs.

Words

abnormal	*adj.*	反常的，变态的
abreast	*adv.*	并肩地，并排地
array	*n.*	排列
	v.	部署，排列
assign	*v.*	分配，指派
	v.	赋值

beyond	*prep.*	在(到)……较远的一边，超过
	adv.	在远处
bug	*n.*	程序缺陷，电脑系统的问题，臭虫
capture	*n.*	捕获
	v.	捕获
consultant	*n.*	顾问，咨询者
Dbx	*n.*	一种调试运行源程序的工具
debug	*v.*	[俗]除错，改正有毛病部分
	n.	调试，调试工具
effort	*n.*	努力，成就
ensure	*v.*	确保，保证
Gdb		一种项目调试器，用来观察程序的执行过程
handy	*adj.*	手边的，容易取得的
issue	*n.*	出版，论点，问题
	v.	使流出，发行(钞票等)
Saber	*n.*	一种 C 调试程序，用来找出程序中的错误，有良好的在线帮助系统
necessary	*n.*	必需，必需品
	adj.	必要的，必然的
pointer	*n.*	指示器
primitive	*adj.*	原始的，简单的
rigorous	*adj.*	严格的
scatter	*v.*	分散
serious	*adj.*	严肃的，认真的，严重的
sophisticate	*n.*	精于……之道的人
	v.	弄复杂，诡辩
stock	*n.*	库存
	adj.	常备的，存货的
subscript	*adj.*	写在下方的
syntactical	*adj.*	[语]依照句法的
threaten	*v.*	恐吓，威胁，可能来临
tip	*n.*	提示

Abbreviations

OLC(Online Learning Center)　　　　在线学习中心

Notes

[1] 例句：This stock answer is intended to give you some tips on how to get started in this area; however, in general, program debugging requires more time and effort

than consultants are usually able to provide.

分析：本句是并列句，however 连接表示转折意义的两个分句。后一个分句中 than consultants are usually able to provide 为 than 引导的比较状语从句, than 在状语从句中充当 provide 的宾语。类似的句子有：John's father seldom spent on his son as much as was necessary…Mark was delighted to see his new flat was larger than I expect…上面的两个句子中，"as"和"than"分别充当了从句的主语和宾语。

译文：这种预存的答案用来给出一些关于在这一领域怎样开始的提示。但总的来说，程序调试需要的时间和努力比咨询者通常所能提供的更多。

[2] 例句：Once you have narrowed down the line where your bug occurs, the next step is to find out the value of your variables at that time.

分析：本句是复合句。这里"once"引导的条件状语从句，意为"一旦……"。主句中不定式短语 to find out the value of your valuables at that time 作 is 的表语。

译文：编程错误存在的范围一旦被缩小了，下一步就是要找出当时的变量值。

[3] 例句：The syntax errors aren't a serious issue during the program design phase since in practice, after designing and testing the design, the program will be implemented in a computer program language and it is at this point that syntax errors become a problem.

分析：本句是复合句。在 since 引导的原因状语从句中，it is at this point that syntax errors become a problem 为强调句。强调句句型结构为：It+be+被强调成分+that/who…。通过这种结构可以强调除谓语动词以外的大多数句子成分。

译文：程序设计阶段，语法错误不是一个严重问题，因为在实际应用中，当设计完成并测试后，要用某一种计算机语言实现程序，这时语法错误才成为需要处理的问题。

Exercises

Ⅰ. Put "true" or "false" in the brackets for the following statements according to the passage.

1. (　) If a program exits abnormally, then there is almost certainly a logical error (a bug) in the program.

2. (　) The first step of debugging is to find the exact line where the program exits.

3. (　) One way of debugging a program is with print statements scattered through the code.

4. (　) Once you have narrowed down the line where your bug occurs, the next step is to find out the value of your variables at that time.

5. (　) In C array subscripts go from 1 to N.

6. (　) Using printfs() is a primitive method of debugging.

7. (　) We can broadly classify the errors as: Syntax errors, Logic errors, Data range and data type errors.

8. (　) Logic errors are a minor problem.

Chapter 5 Programming Language

9. (　) Syntax errors means the algorithm is syntactically correct but doesn't do what is intended.

10. (　) Once launched, the program needs to be maintained.

II. Fill in the blanks according to the passage.

1. When you write a C program, you cannot assign _____ value to a pointer variable.

2. There seems to be always one last _____ or error to be corrected.

3. Syntax errors means that a mistake is made in the _____ used to state the algorithm.

4. Once launched, the program needs to be _____.

5. Program maintenance represents a _____ portion of the total expenditures on application programs.

6. The logic errors are a much more _____ problem.

7. Usually, it is quicker and easier to use a _____.

8. Using printf() is a _____ method of debugging, but sometimes it's the only one that will _____.

9. Program debugging is finding and removing _____ in a program.

10. Bugs are _____ in a program.

III. Translate the following words and expressions into Chinese.

1. necessary
2. check
3. choice
4. tutorial
5. broadly
6. exact
7. platform
8. algorithm
9. implement
10. expenditure

5.4.1 Reading Material

What's Actually Involved in Programming?

What's actually involved in programming—the actual process of writing programs? Here's a quick overview of the process:

- Write a program.
- Compile the program.
- Run the program.
- Debug the program.
- Repeat the whole process until the program is finished.

Let's discuss those steps one by one.

Write a program

Much of a programming language is in English. Programming languages commonly use words like "if", "repeat", "end" and such. Also, they use the familiar mathematical operators like

"+" and "=". It's just a matter of learning the "grammar" of the language: how to say things properly.

So, we said "Write a program". This means: write the steps needed to perform the task, using the programming language you know. You'll do the typing in a programming environment (an application program which lets you write programs, which is an interesting thought in itself).

Incidentally, the stuff you type to create a program is usually called source code, or just code. Programmers also sometimes call programming coding.

Compile the program

In order to use a program, you usually have to compile it first. When you write a program (in a programming language, using a programming environment, as we mentioned a moment ago), it's not yet in a form that the computer can use. This isn't hard to understand, given that computers actually only understand lots of 1s and 0s in long streams. You can't very well write programs using only vast amounts of 1s and 0s, so you write it in a more easily-understood form (a programming language), then you convert it to a form that the computer can actually use. This conversion process is called compiling, or compilation. Not surprisingly, a program called a compiler does the compiling. Compilers are just fancy programs, so they too are written by programmers.

Run the program

Now that you've compiled the program into a form that the computer can use, you want to see if it works: you want to make the computer perform the steps that you specified. This is called running the program, or sometimes executing it. Just the same as how a car isn't of much use if you don't drive it, a program isn't of much use if you don't run it. Your programming environment will allow you to run your program too.

Debug the program

You've probably heard the term "debug" before (it's pronounced just as you might expect: "dee-bug"). It refers to fixing errors and problems with your program. As I'm sure you know, the term came about because the earliest computers were huge building-sized contraptions, and actual real-life insects sometimes flew into the machinery and caused havoc with the circuits and valves. Hence, those first computer engineers had to physically "debug" the computers—they had to scrape the toasted remains of various kinds of flying insects out of the inner workings of their machines. The term became used to describe any kind of problem-solving process in relation to computers, and we use it today to refer purely to fixing errors in our code.

Once again, your programming environment will help you to debug your programs (indeed, you'll often find the picture of an insect shown in your programming environment to indicate debugging). You usually debug your program by stepping through it.

Repeat the whole process until the program is finished

And then you repeat the whole process until you're happy with the program. Programmers are perfectionists—never satisfied until absolutely everything is complete and elegant and powerful and just gorgeous.

And that's the basic process of programming. Note that most programming environments will make a lot of it much easier for you, by doing such things as:

- Warning you about common errors.
- Taking you to the specific bit of code which is causing the compiler to puke.
- Letting you quickly look up documentation on the programming language you're using.
- Letting you just choose to run the program, and compiling it automatically first.
- Colouring parts of your code to make it easier to read (for example, making numbers a different colour from other text).
- And many other things.

So, don't worry too much about the specifics of compiling then running then debugging or whatever. The purpose of this section was mostly to make you aware of the cyclical nature of programming: you write code, test it, fix it, write more, test it, fix, and so on.

Words

commercial	*adj.*	商业的，贸易的
contraption	*n.*	装置
conversion	*n.*	变换，转化
cyclical	*adj*	周期的，循环的
deliberate	*adj.*	故意的，预有准备的
	v.	商讨
enhance	*v.*	提高，增强
gorgeous	*adj.*	宜人的，令人满意的
havoc	*n.*	大破坏，浩劫
	v.	严重破坏
incidentally	*adv.*	附带地，顺便提及
perfectionist	*n.*	完美主义者
puke	*n.*	呕吐
	v.	呕吐
ridiculously	*adv.*	可笑地
scrape	*n.*	刮，擦，困境
	v	刮，擦伤，挖成
slap	*v.*	拍，拍击
	n.	拍
stream	*n.*	溪，流，一串

toast	v.	烤(面包等)，使暖和
trait	n.	显著的特点，特性
tricky	adj.	不易处理的，需要技巧的
whilst	conj.	时时，同时

5.4.2 正文参考译文

程序调试与维护

如果程序非正常退出，很有可能是因为程序中有逻辑错误(或称为 bug)。99%的程序工作是要找出并去除这些 bug。下面是对初学者的一些提示。

首先，再来重复一下关于程序调试的标准 OLC 规则：不要奢望 OLC 会帮助调试程序。这种预存的答案用来给出一些关于在这一领域怎样开始的提示。但总的来说，程序调试需要的时间和努力比咨询者通常所能提供的更多。

第一步是精确地找出程序出错的语句行。实现的方法是在程序代码中分散布置一些打印语句。例如，你可以在源程序中做如下工作：(略)

C 语言中的 fflush()命令使得输出立即被送到显示器上，如果为了调试而使用 printf()，就要用到 fflush()。

编程错误存在的范围一旦被缩小了，下一步就是要找出当时的变量值。你可能会发现某个变量值很奇怪。这时候就要检查你有没有做下面的事情：

- 给指针变量赋予一个整数值；
- 数组的下标越界。(记住，C 数组的下标是从 0 到 N-1，不是从 1 到 N。)

别的错误也会引起 bug，例如要确认循环测试的边界条件使用正确。

其他类型的 bug(程序不终止，错误输出)用同样的方法进行调试。再一次找到第一个错误发生的行，检查变量值。处理一个 bug 后，要重新编译程序，再次运行。如果需要还要再调试。

使用 printf()是调试程序的最原始方法，但有时却是唯一有效的方法。如果你的程序对一个调试器(如 Saber 或 Dbx)来说太大，或者你使用的是非 Athena 平台，就可能没有合适的调试器。通常，使用调试器会更快更容易。Athena 有几个实用的调试工具，Saber 是 C 程序员的工具选择。Gdb 和 Dbx 使用也很方便，并且都适用于 Fortran 和 C。有许多介绍 Saber 和 Dbx 的现成答案，Saber 甚至还有辅导材料。

程序设计中有这样的情况，就是好像永远有一个 bug 或错误有待改正。错误的大体分类如下：

- 语法错误——这一类错误出现在表示算法的语言中。
- 逻辑错误——表示算法的语法是正确的，但得不到预期结果。
- 数据范围和数据类型错误——算法的语法和逻辑都正确，但是有错误数据类型或超出范围的数据值对程序造成威胁。

程序设计阶段，语法错误并不是一个严重问题，因为在实际应用中，当设计完成并测试后，要用某一种计算机语言实现程序，这时语法错误才成为需要处理的问题。但即使这样，语法错误也是小问题，因为在程序组建阶段会发现错误。只有清除语法错误后才能组建程序。

逻辑错误是较为严重的问题，因为没有办法消除逻辑错误，除非彻底测试程序设计。

数据错误也是严重错误，有时比逻辑错误更难处理。

程序投入使用后，常需要维护。程序维护就是指根据机构的需要或硬件、软件改变的情况不断地更新程序。基于要实现的维护任务，程序管理员需要决定后续经济支持和人员的需要及配备情况。程序维护是应用程序总体花费中的主要组成部分。

5.4.3 阅读材料参考译文

编程究竟包括什么？

编程究竟包括什么呢，也就是说，编写程序的实际过程是什么？这里快速看一下该过程：
- 写程序。
- 编译程序。
- 运行该程序。
- 调试程序。
- 重复整个过程，直到该程序完成。

下面一一讨论这些步骤。

写程序

许多编程语言是英语。编程语言通常使用"if"、"repeat"、"end"等之类的词语。同时，它们利用熟悉的数学运算符，如"+"和"="。这只是学习语言"语法"的一个问题：如何正确地说出事物。

所以，我们说"写程序"。这意味着：使用所掌握的编程语言，编写完成任务的必要步骤。您将在一个编程环境里输入程序(一个可以编写程序的应用程序，这本身是一个非常有趣的想法)。

顺便说一下，为形成一个程序而输入的材料通常称为源代码，或者只是称为代码。程序员有时把编程叫做编码。

编译程序

为了使用某个程序，通常首先要编译该程序。当写完一个程序(用一种编程语言、使用一个编程软件，正如刚才提到的)，这一程序还未形成电脑可以使用的形式。这是不难理解的，因为电脑实际上只能识别大量长串的 1 和 0。只使用大量的 1 和 0 不能很好地写程序，所以要用更容易理解的形式(一种编程语言)编写程序，然后将其转换为计算机可以实际使用的形式。这个转换过程称为编译。使用称为编译器的程序进行编译，这不足为奇。编译器只是奇特的程序，所以它们也是程序员编写的。

运行程序

既然已经把程序编译成计算机可以使用的格式，下面就看看它是否可以工作：使计算机执行指定的步骤。这称为运行程序，有时称为执行程序。一辆汽车如果不行驶就没有多少用处，同样，一个程序如果不运行也就没有什么用处。编程环境也可以允许运行程序。

调试程序

大多数人以前可能听到过"调试"这个术语(它的读音正如可能期望的一样:"dee-bug")。它指的是修改程序的错误和问题。正如大家所知,该术语的来历是,最早的计算机的装置体积非常巨大,有时会有昆虫飞入机器,造成电路和阀门损坏。那些早期的计算机工程师们不得不亲自为电脑"捉虫",他们需要把各类飞行昆虫的烤煳了的遗骸从机器内部刮出来。后来人们用这个词描述与电脑有关的任何一种解决问题的过程,今天我们用它单纯指修改代码中的错误。

同时,编程环境将帮助调试程序(实际上,我们经常会发现编程环境中显示昆虫图片——表示需要调试)。一般情况下,可以通过逐步跟踪程序进行调试。

重复整个过程,直到编程完成

然后重复整个过程,直到对程序满意。程序员多是完美主义者,永不满足,直到一切都绝对完美。

这是编程的基本过程。注意大多数编程环境下通过做以下事情使许多事情变得更容易。

- 给出常见错误的警告。
- 找到造成编译失败的具体代码。
- 快速查找所用编程语言的有关文件。
- 选择运行该程序,并首先自动编译该程序。
- 改变部分代码的颜色使其更易于阅读(例如,使数据出现与其他文字不同的颜色)。
- 其他的许多事情。

所以,不要太担心编译、运行、调试以及其他任何事情的具体细节。本节的目的主要是帮助您了解编程的周期性:写代码,测试、修正代码,再写代码、测试、修正代码,以此循环。

Chapter 6　Information Security

6.1　Concept of Information Security

The issue of information security and data privacy is assuming tremendous importance among global organizations, particularly in an environment marked by computer viruses and terrorist attacks, hackings and destruction of vital data owing to natural disasters. [1] When it comes to information security, most companies fall somewhere between two extreme boundaries: complete access and complete security. A completely secure computer is one that is not connected to any network and physically unreachable by anyone. A computer like this is unusable and does not serve much of a practical purpose. On the other hand, a computer with complete access is very easy to use, requiring no passwords or authorization to provide any information. [2] Unfortunately, having a computer with complete access is also not practical because it would expose every bit of information publicly, from customer records to financial documents. Obviously, there is a middle ground—this is the art of information security.

The concept of information security is centered on the following components:
- Integrity: gathering and maintaining accurate information and avoiding malicious modification.
- Availability: providing access to the information when and where desired.
- Confidentiality: avoiding disclosure to unauthorized or unwanted persons.

For an information system to be secure, it must have a number of properties:

service integrity. [3]This is a property of an information system whereby its availability, reliability, completeness and promptness are assured.

data integrity. This is a property whereby records are authentic, reliable, complete, unaltered and useable, and the processes that operate on them are reliable, compliant with regulatory requirements, comprehensive, systematic, and prevent unauthorized access, destruction, alteration or removal of records. These requirements apply to machine-readable databases, files and archives, and to manual records.

data secrecy. This is a property of an information system whereby information is available only to those people authorized to receive it. Many sources discuss secrecy as though it was only an issue during the transmission of data; but it is just as vital in the context of data storage and data use.

authentication. Authentication is a property of an information system whereby assertions are checked. Forms of assertion that are subjected to authentication include:
- "**data** authentication", whereby captured data's authenticity, accuracy, timeliness, completeness and other quality aspects are checked.

- "**identity** authentication", whereby an entity's claim as to its identity is checked. This applies to all of the following:
  - the identity of a person;
  - the identity of an organizational entity;
  - the identity of a software agent; and
  - the identity of a device.
- "**attribute** authentication", whereby an entity's claim to have a particular attribute is checked, typically by inspecting a "credential". Of especial relevance in advanced electronic communications is claim of being an authorized agent, i.e. an assertion by a person, a software agent or a device to represent an organization or a person.

Non-repudiation. This is a property of an information system whereby an entity is unable to convincingly deny an action it has taken.

There is a strong tendency in the information systems security literature to focus on the security of data communications. But security is important throughout the information life-cycle, i.e. during the collection, storage, processing, use and disclosure phases, as well as transmission. Each of the properties of a secure system identified above needs to be applied to all of the information life-cycle phases.

Words

authorization	n.	授权，认可
availability	n.	可用性，有效性
boundary	n.	分界线
component	n.	成分
	adj.	组成的，构成的
confidentiality	n.	机密性
convincingly	adv.	有说服力地
credential	n.	凭证
entity	n.	实体
financial	adj.	财政的，金融的
hacker	n.	电脑黑客
integrity	n.	完整性
organizational	adj.	组织的，机构的
owing	adj.	(~to)由于，应归功于
plug in	v.	插上电源
promptness	n.	敏捷，机敏
relevance	n.	中肯，适当
repudiation	n.	批判
secrecy	n.	秘密，保密
terrorist	n.	恐怖分子

Chapter 6 Information Security

unauthorized	adj.	未被授权的，未经认可的
unreachable	adj.	不能达到的
virus	n.	病毒
vital	adj	重大的，至关重要的，所必需的
whereby	adv.	由此，赖以

Notes

[1] 例句：When it comes to information security, most companies fall somewhere between two extreme boundaries: complete access and complete security.

分析：本句是复合句。when it comes to sth/doing sth 意为"当涉及(做)某事物的情况、事情和问题时"，to 在此作介词用，其后接名词或动名词。例如：

When it comes to studying hard, Gary burns the midnight oil almost every night.
说到用功学习，加里几乎每晚都熬通宵。

译文：涉及信息安全，很多公司会处于两个极端：完全开放和完全安全。

[2] 例句：Unfortunately, having a computer with complete access is also not practical because it would expose every bit of information publicly, from customer records to financial documents.

分析：本句是复合句。句中 unfortunately 为评注性状语。评注性状语不是修饰谓语或谓语动词，而是对整个句子进行说明或解释，表明说话人对问题的看法或态度，通常位于句首并用逗号与句子隔开。

译文：遗憾的是，拥有完全访问的计算机也是不实际的。因为它会公开暴露每一点信息，从客户记录到财政文件。

[3] 例句：This is a property of an information system whereby its availability, reliability, completeness and promptness are assured.

分析：本句是复合句。句中 whereby 为关系副词，引导定语从句，相当于 by which，意为"靠那个，凭那个，借以"。

译文：这是信息系统的一个特性，以保证信息系统具有有效性、可靠性、完整性和敏捷性。

Exercises

Ⅰ. Put "true" or "false" in the brackets for the following statements according to the passage.

1. (   ) Complete access and complete security are good for information security.

2. (   ) A completely secure computer is unusable and does not serve much of a practical purpose.

3. (   ) Having a computer with complete access is practical.

4. (   ) Integrity means gathering and maintaining accurate information and avoiding malicious modification.

5. (   ) Confidentiality means providing access to the information when and where desired.

6. (   ) Availability means avoiding disclosure to unauthorized or unwanted persons.

7. (　) Data secrecy is a property of an information system whereby information is available only to those people authorized to receive it.

8. (　) Authentication is a property of an information system whereby assertions are checked.

9. (　) Security is important throughout the information life-cycle.

10. (　) During transmission, information security is not important.

II. Fill in the blanks according to the passage.

1. A completely _____ computer is one that is not connected to any network and physically unreachable by anyone.

2. A computer with complete access is very _____ to use, requiring no _____ or authorization to provide any information.

3. The concept of information security is centered on the following components: _____, _____, and confidentiality.

4. Service integrity is a property of an information system whereby its _____, _____, completeness and promptness are assured.

5. Data secrecy is a property of an information system whereby information is available only to those people _____ to receive it.

6. Forms of assertion that are subjected to authentication include: data _____, identity authentication, and _____ authentication.

7. Non-repudiation is a property of an information system whereby an entity is unable to convincingly _____ an action it has taken.

8. There is a strong tendency in the information systems security literature to focus on the security of data communications.

9. Security is important throughout the _____.

10. There is a _____ between complete access and complete security.

III. Translate the following words and expressions into Chinese.

1. importance
2. environment
3. destruction
4. disaster
5. unreachable
6. unusable
7. maintain
8. process
9. database
10. communication

6.1.1　Reading Material

Information Security System

The information security system is an integral part of the national security system. The main functions of the information security system are:

- Assessing the state of information security in the country, identifying and forecasting internal and external threats to information security, drafting an information security doctrine.

Chapter 6 Information Security

- Developing a comprehensive system of legal, administrative, economic, technical and other measures and methods aimed at ensuring information security.
- Coordinating and monitoring the work of information security entities.
- Protecting information security entities against incomplete, inaccurate and distorted information and against exposure to information damaging to their life and health.
- Protecting protected information.
- Counteracting technical intelligence services.
- Developing and perfecting an information infrastructure, an information technology industry, systems, means and services.
- Organizing scientific research, developing and implementation of scientific, scientific-technical programs in the field of information security; licensing the activities of corporations and individual entrepreneurs in the field of information security.
- Certifying information systems and means, assessing and rating the compliance of information facilities with information protection requirements.
- State inspection in the field of information security.
- Creating conditions for preserving and developing intellectual potential in the information sphere.
- Preventing, identifying and suppressing offences which are aimed at hurting the rights and freedoms of corporations and individuals in the information sphere, prosecuting and trying in court perpetrators of crimes in the information sphere.
- Carrying out international cooperation in the sphere of information security.

Words

administrative	*adj.*	管理的，行政的
assess	*v.*	估定，评定
certify	*v.*	证明，保证
compliance	*n.*	依从，顺从
coordinate	*n.*	坐标(用复数)
	adj.	同等的，并列的
	v.	调整
counteract	*v.*	抵消，中和，阻碍
doctrine	*n.*	教条，学说
draft	*v.*	起草
entrepreneur	*n.*	[法]企业家，主办人
forecast	*v.*	预测
identify	*v.*	识别，鉴别，确定
inspection	*n.*	检查，视察
integral	*adj.*	完整的，构成整体所需要的

	n.	部分
offence	n.	犯罪，冒犯，[军] 攻击
perpetrator	n.	犯罪者，作恶者
preserve	v.	保存，保留
prosecute	v.	实行，起诉，告发，作检察官
rating	n.	等级，级别，额定
sphere	n.	领域，方面
suppress	v.	抑制，查禁
threat	n.	恐吓，威胁

6.1.2　正文参考译文

信息安全的概念

　　信息安全和数据保密问题在全世界的机构中都非常重要，尤其在一个以计算机病毒、恐怖分子、黑客攻击以及自然灾害造成重要数据破坏为特点的环境中。涉及信息安全，很多公司会处于两个极端：完全开放和完全安全。完全安全的计算机是不连接任何网络的，也不让任何人接触。像这样的计算机是不能用的，也没有多少实际用途。另一方面，完全开放的计算机容易使用，提供任何信息不需要密码和权限。遗憾的是，拥有完全访问的计算机也是不实际的，因为它会公开暴露每一点信息，从客户记录到财政文件。显然其中有一个中间区域——这就是信息安全的技术。

　　信息安全的概念集中于下面几部分：
- 完整性：收集和维护正确的信息并且避免恶意破坏。
- 可用性：随时随地根据需要提供对信息的访问。
- 机密性：避免向未授权或不必要的人泄漏信息。

　　安全的信息系统需要具有如下的特性：

　　服务完整性。信息系统的这一特性保证信息系统具有有效性、可靠性、完整性和敏捷性。

　　数据完整性。信息系统的这一特性保证记录可信、可靠、完全、无改变、而且可用。对记录进行的操作可靠、适应调整要求、全面、系统，而且能阻止未授权者访问、破坏、改变或删除纪录。这些要求适用于可用计算机处理的数据库、文件以及档案，也适用于人工记录。

　　数据保密。信息系统的这一特性使得信息只对有授权的人可用。很多资料把安全仅作为数据传输中的问题讨论，但是在数据存储和使用环境中安全问题同样重要。

　　鉴定。信息系统这一特性保证行为可以被检查。可以鉴定的行为方式包括：
- 数据鉴定，检查已得数据的真实性、精确性、时效性、完全性以及其他方面的性能。
- 身份鉴定，鉴定实体宣称的身份。可应用于鉴定人员的身份、机构实体的身份、软件代理的身份和设备的身份。
- 属性鉴定，通常通过检查"凭证"，鉴定一个实体声称自己具有的特定属性。先进的电子通信领域内最为相关的是授权代理的声明，也就是个人、软件代理商或设备代表某个机构或个人所做的声明。

　　不可否认性。信息系统这一特性使实体不能否认所执行过的行为。

在有关信息系统安全的文献中,多数文献强调关注数据通信中的安全,但是在整个信息处理周期中安全都是很重要的,也就是说,安全在信息的收集、存储、处理、使用和公布阶段,与在传输阶段同样重要。上述安全系统的特性需要应用于信息处理周期的各阶段。

6.1.3 阅读材料参考译文

信息安全体系

信息安全体系是国家安全体系的组成部分,其主要功能有:
- 评估国家信息安全的状况,查明和预测信息安全的内部和外部威胁,起草信息安全规则。
- 改进涉及法律、行政、经济、技术和其他措施和办法的全面体系,以确保信息安全。
- 协调和监督信息安全实体的工作。
- 保护信息安全实体,避免不完整、不准确和歪曲的信息,避免接触破坏信息安全实体生命和健康的信息。
- 保护受保护的资料。
- 对抗技术情报服务。
- 发展和完善信息基础设施、信息技术行业、体系、方法和服务。
- 组织科研,开发、实施信息安全领域的科学技术方案;特许信息安全领域企业和个体企业家的活动。
- 认证信息体系和方法,对具有信息保护要求的信息设施的遵守情况进行评估和分级。
- 信息安全领域中的状态检测。
- 创造条件以维护和发展信息领域中的智力潜能。
- 预防、查明和制止信息领域内损害企业和个人权利及自由的违法行为,起诉和指控信息领域内的罪犯。
- 开展信息安全领域的国际合作。

6.2 Computer Viruses

[1] Just as human viruses invade a living cell and then turn it into a factory for manufacturing viruses, computer viruses are small programs that replicate by attaching a copy of themselves to another program. Once attached to the host program, the virus then look for other programs to "infect". In this way, the virus can spread quickly throughout a hard disk or an entire organization if it infects a LAN (Local Area Network) or a multi-users system.

[2] Skillfully written virus can infect and multiply for weeks or months without being detected. During that time, system backups duplicate the viruses, or copies of data or programs made and passed to other systems to infect. At some point—determined by how the virus was programmed—the virus attacks. [3]The timing of the attack can be linked to a number of situations, including: a certain time or date; the presence of a particular user ID; the use or presence of a particular file; the security privilege level of the user; and the number of times of a

file is used.

Likewise, the mode of attack varies, so-called "being" viruses might simply display a message, like the one that infected IBM's main computer system last Christmas with a season's greeting.

Malignant viruses, on the other hand, are designed to damage your system. One common attack is to wipe out data, to delete files, or to perform a format of disk.

There are four main types of viruses: shell, intrusive, operating system, and source code.

- Shell viruses wrap themselves around a host and do not modify the original program. Shell program are easy to write, which is why about half of all viruses are of this type. In addition, shell viruses are easy for programs like Data Physician to remove.
- Intrusive viruses invade an existing program and actually insert a portion of themselves into the host program. Intrusive viruses are hard to write and difficult to remove without damaging the host file.

Shell and intrusive viruses most commonly attack executable program file—those with. COM or. EXE extension—although data are also at some risk.

- Operating system viruses work by replacing parts of operating system with their own logic. [4]Very difficult to write, these viruses have the ability, once booted up, to take total control of your system. According to Digital Dispatch, known versions of operating system viruses have hidden large amounts of attack logic in falsely marked bad disk sectors. Others install RAM-resident programs or device drivers to perform infection or attack functions invisibly from memory.
- Source code viruses are intrusive programs that are inserted into a source program such as those written in Pascal prior to the program being compiled. These are the least common viruses because they are not only hard to write, but also have a limited number of hosts compared to the other types.

New computer viruses are written all the time, and it's important to understand how your system can be exposed to them and what can do to protect your computer. Follow the suggestions listed below to substantially decrease the danger of infecting your computer system with a potentially dangerous computer virus.

- Be very cautious about inserting disks from unknown sources into your computer.
- Always scan the disk's files before operating any of them.
- Only download Internet files from reputable sites.
- Do not open e-mail attachments (especially executable files) from strangers.
- Purchase, install, and use an anti-virus software program. The program you choose must provide three functions:
    - Detection.
    - Prevention.
    - Removal.

As new viruses are created everyday, upgrade your anti-virus software regularly.

Chapter 6 Information Security

Words

attach	v.	附加，隶属
anti-virus	a.	防病毒的，抗病毒的
backup	n.	后备，备份
cautious	adj.	细心的，谨慎的
detect	v.	发觉，发现，侦测
duplicate	adj.	复制的
driver	n.	驱动程序
detection	n.	发觉，侦察
infect	v.	传染，侵染，感染
invade	v.	侵入，侵犯
intrusive	adj.	侵入的，闯入的
ID	n.	身份标识
manufacture	v.	加工，制造
mode	n.	方式，样式
multiply	v.	增加，成倍地增加
malignant	adj.	恶意的，有害的
privilege	n.	特权，特许
potentially	adv.	潜在地
purchase	n.	购买，购置
prevention	n.	预防，防止
replicate	v.	复制，重复
reputable	adj.	声誉好的，可尊敬的
resident	adj.	常驻的，居留的
removal	n.	移动，除掉
regularly	adv.	有规律地，有规则地
situation	n.	位置，地位，情况，局面
shell	n.	外壳
substantially	adv.	潜在地
upgrade	n./v.	升级，使升级
wipe	v.	擦去，去除
wrap	v.	包，裹，隐藏，伪装

Notes

[1] 例句：Just as human viruses invade a living cell and then turn it into a factory for manufacturing viruses, computer viruses are small programs that replicate by attaching a copy of themselves to another program.

分析：本句是复合句，just as 引导的是比较状语从句。主句中又包含定语从句 that

replicate by attaching a copy of themselves to another program，修饰 programs。

译文：正如人的病毒侵害活细胞，然后将它变成制造病毒的工厂一样，计算机病毒是小程序，通过将该程序本身的副本附加到另一个程序上来进行复制。

[2] 例句：Skillfully written virus can infect and multiply for weeks or months without being detected.

分析：本句是简单句。skillful written 是过去分词短语作定语，修饰 viruses。

译文：巧妙地编写的病毒会在数周或数月内进行传染并倍增，而不会被人们发现。

[3] 例句：The timing of the attack can be linked to a number of situations, including: a certain time or date; the presence of a particular user ID; the use or presence of a particular file; the security privilege level of the user; and the number of times a file is used.

分析：本句是复合句。a file is used 是一个定语从句，修饰 the number of times。

译文：攻击的时刻可能与许多情况有关，包括：某一特定时间或日期、特定用户身份的出现、特定文件的使用或出现、用户的安全优先级别以及某一文件被使用的次数。

[4] 例句：Very difficult to write, these viruses have the ability, once booted up, to take total control of your system.

分析：本句是复合句。从句 once booted up 是 once they are booted up 的省略形式，作插入语。

译文：这种病毒很难编写，而一旦被引导，它们就能够完全控制整个系统。

Exercises

Ⅰ. Put "true" or "false" in the brackets for the following statements according to the passage.

1. ( ) Computer viruses can replicate copies of themselves attaching to another program.

2. ( ) There are four main types of viruses: shell, intrusive, operating system, and source code.

3. ( ) Intrusive viruses wrap themselves around a host and do not modify the original program.

4. ( ) In order to protect your computer you should only download Internet files from reputable sites.

5. ( ) Be very cautious about inserting disks from unknown sources into your computer.

6. ( ) We may download Internet files from any site in order to decrease the danger of infecting your computer system.

7. ( ) Do not open e-mail attachments (especially executable files) from strangers.

8. ( ) Intrusive viruses are hard to write and difficult to remove without damaging the host file.

9. ( ) Operating system viruses work by replacing parts of operating system with their own logic.

10. ( ) Source code viruses are intrusive programs that are inserted into a source program.

Chapter 6 Information Security

II. Fill in the blanks according to the passage.

1. Once attached to the host program, the virus then look for other programs to "_____".
2. The timing of the attack can be linked to a number of situations, including: a certain time or date; the presence of a particular_____; the use or presence of a _____; the security privilege level of the user; and the number of times of a file is used.
3. There are four main types of viruses: _____, intrusive, _____, and source code.
4. As new viruses are created everyday, upgrade your _____ regularly.
5. Computer viruses are small program that_____ by attaching a copy of themselves to another program.
6. Skillfully written virus can infect and multiply for weeks or months without _____.
7. One common attack is to _____ data, to delete files, or to perform a format of disk.
8. Intrusive viruses _____ an existing program and actually insert a portion of themselves into the host program.
9. _____ viruses are intrusive programs that are inserted into a source program.
10. As new viruses are created everyday, _____ your anti-virus software regularly.

III. Translate words and expressions into Chinese.

1. computer virus
2. host program
3. LAN
4. a particular user ID
5. security privilege level
6. shell virus
7. intrusive virus
8. operating system virus
9. reputable site
10. anti-virus software program

6.2.1 Reading Material

Backdoors

Backdoor programs are typically more dangerous than computer viruses, as they can be used by an intruder to take control of a PC and potentially gain access to an entire network.

Backdoor programs, also referred to as Trojan horses, are typically sent as attachments to e-mails with innocent-looking file names, tricking users into installing them. They often enable remote users to listen in on conversations using the host computer's microphone, or even see through its video camera if it has one. Back Orifice (BO) 2000 is a backdoor program designed for malicious use. Its main purpose is to maintain unauthorized control over another machine for reconfiguration and data collection. It takes the form of a client/server application that can remotely control a machine without the user's knowledge to gather information, perform system commands, reconfigure machines and redirect network traffic.

With BO an intruder has to know the user's IP address to connect, or could scan an entire network looking for the victim. Once connected, the intruder can send requests to the BO 2000 server program, which performs the actions the intruder specifies on the victim's computer, sending back the results.

BO is installed on the server machine simply through the execution of the server application. This executable file is originally named bo2k.exe, but it can be renamed. The configuration wizard will step through the various configuration settings, including the server file (the executable), the network protocol, port number, encryption, and password. Once this process is complete, running bo2kgui.exe executes the user interface for BO.

It is very difficult to detect BO, because it is so highly configurable. In addition, backdoor programs are multi-dimensional, so several detection methods are recommended to achieve maximum protection and awareness of the installation of BO 2000 on a machine or series of machines on a network.

We recommend coupling the use of an updated version of anti-virus software to detect which machines on the network have BO installed—and intrusion detection software to identify attacks over the network.

Users are urged to follow three important precautions:

Do not accept files from Internet chat systems.

If you are connected to the Internet, do not enable network sharing without proper security in place.

Do not open e-mail attachments: never run any executable files sent to you (.exe files or .zip files with a.exe in them). It is safer if these are run through a virus checker first, but they could be new backdoor programs or viruses that a virus scanner will not detect. It is safe to open Word documents and Excel spreadsheets if the Microsoft Auto-Run feature is turned off. Allowing macros to run automatically can spread e-mail viruses such as Melissa. Many people send each other animations in e-mail: it is easy to put a backdoor program into one of these and users cannot tell when they infect their computers with Back Orifice 2000.

Words

animation	n.	动画(制作)
attachment	n.	附件
backdoor	n.	后门
configurable	adj.	可配置的
detection	n.	察觉，侦查，探测
encryption	n.	编密码
innocent	adj.	清白的，无害的，无知的
intruder	n.	入侵者
intrusion	n.	侵入，闯入
malicious	adj.	怀恶意的，预谋的
potentially	adv.	潜在地
precaution	n.	预防，警惕
reconfiguration	n.	重新配置，重新组合

| redirect | *vt.* | 使改变方向 |
| urge | *v.* | 催促，强烈要求 |

6.2.2 正文参考译文

计算机病毒

正如人的病毒侵害活细胞，然后将它变成制造病毒的工厂一样，计算机病毒是小程序，通过将该程序本身的副本附加到另一个程序上来进行复制。病毒一旦附加在主程序上，就会寻找其他程序进行"感染"。以这种方式，病毒迅速传播到整个硬盘，如果病毒感染了局域网或多用户系统，则会迅速传染整个机构。

巧妙地编写的病毒会在数周或数月内进行传染并倍增，而不会被人们发现。在这一段时间内，系统的备份复制此病毒，或者数据或程序的副本被制造出来并传递到其他系统进行传染。病毒在某一点开始攻击，"某一点"取决于该病毒的编程方式。攻击的时刻可能与许多情况有关，例如：某一特定时间或日期、特定用户身份的出现、特定文件的使用或出现、用户的安全优先级别以及某一文件被使用的次数。

同样，攻击模式变化多端。所谓"生物"病毒，可能只显示一条信息；就像前几年圣诞节曾经传染了 IBM 主计算机系统的病毒，它显示的信息是节日问候。

另一方面，致命的病毒是被设计用来破坏计算机系统的。一种常见的攻击方式是擦除数据、删除文件或对硬盘进行格式化。

共有四种主要类型的病毒：壳型、侵入型、操作系统型和源代码型。

- **壳型病毒**将自己捆绑在主程序上，并不修改原始程序。壳型程序易于编写，所以大约一半的病毒均属于这种类型。此外，对于像"数据医生"这样的程序来说，壳型病毒是易于被清除的。
- **侵入型病毒**侵害现有的程序。实际上，它们把自己的一部分插入到主程序中。侵入型病毒难以编写，而且在不破坏主文件的情况下难以清除。

壳型和侵入型病毒最常攻击的是可执行文件(即带有.COM 或.EXE 扩展名的文件)，但其实数据文件也有某种危险。

- **操作系统型病毒**用自己的逻辑替换操作系统(OS)的某些部分。这种病毒很难编写，而一旦被引导，它们就能够完全控制整个系统。根据 Digital Dispatch 的说法，已知的操作系统型病毒版本在虚假标记为坏的磁盘扇区内隐藏了大量的攻击逻辑。这一类型的其他病毒安装 RAM 驻留程序或设备驱动程序，以便不被觉察，就从内存执行感染或攻击功能。
- **源代码型病毒**是一种侵入式程序，在源程序被编译之前，这种程序被插入到例如用 Pascal 编写的源程序中。这种病毒最不常见，这是因为它们不仅难以编写，而且与其他类型的病毒相比，其宿主的数量很有限。

一直都有人在编写新病毒，了解系统是如何暴露给病毒的以及做什么能够保护计算机很重要。下面的建议可以帮您大大减少某种潜在的计算机病毒感染给计算机系统造成的危险。

- 当来源不明的磁盘插入计算机时，要非常小心。
- 在打开任何文件之前，先扫描一下磁盘文件。

- 仅从信誉好的网站下载文件。
- 不打开来自陌生人的电子邮件的附件(特别是可执行的文件)。
- 购买、安装及使用防病毒软件程序,所选择的这种程序必须提供三种功能:检测、预防、清除。

因为每天都有新病毒产生,所以要定期升级防病毒软件。

6.2.3 阅读材料参考译文

后门程序

后门程序一般比计算机病毒更加危险,因为入侵者可通过它来控制一台电脑,继而有可能进入整个网络。

后门程序,又称为特洛伊木马,通常作为电子邮件的附件发送。它们的文件名看起来无害,以诱骗用户安装。这类程序能够使远程用户通过主机麦克风进行窃听,甚至可透过其摄像机(只要有的话)来偷看。Back Orifice 2000 (简称 BO)便是这样一种恶意后门程序。其主要目的就是越权控制另一台电脑,以进行重新配置和数据搜集。它是一种客户-服务器应用程序,在用户不知道的情况下对其电脑进行远程控制,以搜集信息、执行系统命令、重新配置电脑和改变网络传输路径。

使用 BO 时,入侵者必须知道用户 IP 地址才能连接,或者通过扫描整个网络来寻找攻击对象。一旦连接成功,入侵者就可以向 BO2000 服务器程序提出请求,该程序便在受害人电脑上做出指定的操作,再把结果发回给入侵者。

只要执行服务应用程序,BO 便被安装到服务器上。这种可执行文件的原始文件名为 bo2k.exe,但它可以被重新命名。配置向导将按步骤进行各种配置设置,包括服务器文件(可执行)、网络协议、端口号、加密和口令。整个过程一旦完成,运行 bo2kgui.exe 便会出现 BO 用户界面。

由于 BO 程序可配置性极高,对其进行检测就非常困难。此外,后门程序是多维的。因此,要使用多种检测方法来最大程度地防范并知晓 BO2000 是否已装上了你的机器或网络上的其他机器。

建议使用最新的杀毒软件来检测网络上哪些电脑已被安装了 BO,同时使用入侵检测软件来识别网上攻击。

强烈要求用户遵循以下三种重要预防措施:

不从因特网聊天系统上接收文件。

如果你连接了因特网,在没有适当的安全措施时,不要启用网络共享功能。

不要打开电子邮件附件:千万不要运行别人发来的可执行文件(例如后缀为 EXE 的文件,或包含有 EXE 文件的 ZIP 文件)。较安全的做法是先用查毒软件检测,但如果它们可能是新型的非法入侵程序或病毒,查毒扫描软件就可能检测不到。如果关闭了微软(Microsoft)的自动运行(宏)功能,打开 Word 文档和 Excel 电子表格就没什么安全问题。任由宏自动运行,将会传播诸如 Melissa(美丽莎)这样的电子邮件病毒。不少人相互在电子邮件中发送动画,后门程序很容易藏身其中,因此用户就很难说清 BO2000 何时会感染他们的计算机。

6.3 Internet Security

In recent years, Internet changes our life a lot. We use e-mail and Internet phone to talk with our friends, we get up-to-date information through web and we do shopping in the cyber-market. Internet has many advantages over traditional communication channels, e.g. it's cost effective, it delivers information fast and it is not restricted by time and place. [1]The more people use Internet, the more concerns about Internet security.

In person-to-person community, security is based on physical cues. To name but a few, we use our signature to authenticate ourselves; we seal letters to prevent others inspection and modification; we receive receipt with the shop's chop to make sure we paid; we get information from a reliable source. But in the Internet society, no such physical cue is available. There are two areas that we show concern about in Internet communication. The first one is secrecy—how do we ensure no one reads the data during its transmission? The second one is authentication—how can we be sure that the identity of someone claiming "who it is".

Encryption is the way to solve the data security problem. In real life, if Tom wants to talk with Mary secretly, he can choose a room with nobody there and talk with Mary quietly, or he can talk with Mary using codes understandable by Tom and Mary only. We take the second approach—encryption—to transmit data through Internet. There are two kinds of encryption techniques—symmetric key encryption and asymmetric key encryption.

For symmetric key encryption, both parties should have a consensus about a secret encryption key. When A wants to send a message to B, A uses the secret key to encrypt the message. After receiving the encrypted message, B uses the same (or derived) secret key to encrypt the message. The advantage of using symmetric key encryption lies in its fast encryption and decryption processes (when compared with asymmetric key encryption at the same security level). The disadvantages are, first, the encryption key must be exchanged between two parties in a secure way before sending secret messages. Secondly, we must use different keys with different parties. For example, if A communicates with B, C, D and E, A should use 4 different keys. Otherwise, B will know what A and C as well as A and D have been talking about. The drawbacks of symmetric key encryption make it unsuitable to be used in the Internet, because it's difficult to find a secure way to exchange the encryption key.

For asymmetric key encryption, there is a pair of keys for each party: a public key and a private key. The public key is freely available to the public, but only the key owner gets hold of the private key. Messages encrypted by a public key can only be decrypted by its corresponding private key, and vice versa. When A sends a message to B, A first gets B's public key to encrypt the message and sends it to B. After receiving the message, B uses his private key to decrypt the message. The advantage comes in the public key freely available to the public, hence free from any key exchange problem. The disadvantage is the slow encryption and decryption process. Asymmetric key cryptography seems to attain secrecy in data transmission, but the authentication problem still exists. Consider the following scenario: when A sends a message to B, A gets B's public key from the Internet—but how can A know the public key obtained actually belongs to B? Digital certificate emerges to solve this problem.

Digital certificate is an identity card counterpart in the computer society. When a person wants to get a digital certificate, he generates his own key pair, gives the public key as well as some proof of his identification to the Certificate Authority (CA). CA will check the person's identification to assure the identity of the applicant. [2]If the applicant is really the one "who claims to be", CA will issue a digital certificate, with the applicant's name, e-mail address and the applicant's public key, which is also signed digitally with the CA's private key. When A wants to send B a message, instead of getting B's public key, A now has to get B's digital certificate. A first checks the certificate authority's signature with the CA's public key to make sure it's a trustworthy certificate. Then A obtain B's public key from the certificate, and uses it to encrypt the message and sends it to B.

Authentication is an important part of everyday life. The lack of strong authentication has inhibited the development of electronic commerce. It is still necessary for contracts, legal documents and official letters to be produced on paper. Strong authentication is then, a key requirement if the Internet is to be used for electronic commerce. Strong authentication is generally based on modern equivalents of the one time pad. For example, tokens are used in place of one-time pads and are stored on smart cards or disks.

[3] Many people pay great amounts of lip service to security, but do not want to be bothered with it when it gets in their way. It's important to build systems and networks in such a way that the user is not constantly reminded of the security system around him. Users who find security policies and systems too restrictive will find ways around them. Security is everybody's business, and only with everyone's cooperation, an intelligent policy, and consistent practices, will it be achievable.

Words

authenticate	v.	鉴别
authentication	n.	证明，鉴定
applicant	n.	申请者，请求者
cyber-market	n.	网上商店
cryptography	n.	密码系统，密码术
certificate	n.	证书
	v.	发给证明书
counterpart	n.	副本，配对物
cooperation	n.	合作，协作
decrypt	v.	解密，解释明白
encryption	n.	加密术，密码术
identification	n.	辨认，鉴定，证明

receipt	n.	收据，收条
signature	n.	签名
scenario	n.	情况说明，游戏的关或是某一特定情节
token	n.	令牌
trustworthy	adj.	可信赖的
up-to-date	adj.	最近的，当代的

phrases

get in the way	妨碍
lip service	说得好听的话，空口的应酬话
public key	公开密钥

Abbreviations

| CA(Certificate Authority) | 证书授权机构 |

Notes

[1] 例句：The more people use Internet, the more concerns about Internet security.
分析：这是"the+形容词比较级，the +形容词比较级"句型，表示"越……越……"。
译文：使用 Internet 的人越多，对 Internet 安全的关注就越多。

[2] 例句：If the applicant is really the one "who claims to be", CA will issue a digital certificate, with the applicant's name, E-mail address and the applicant's public key, which is also signed digitally with the CA's private key.
分析：CA 指"证书授权机构"，是可信任的第三方，它保证数字证书的有效性。CA 负责注册、颁发证书，并在证书包含的信息变得无效后收回证书。
译文：如果申请人确如自己所声称的，证书授权机构将授予带有申请人姓名、电子邮件地址和申请人公钥的数字证书，并且该数字证书由证书授权机构用其私有密钥进行数字签名。

[3] 例句：Many people pay great amounts of lip service to security, but do not want to be bothered with it when it gets in their way.
分析：lip service 意思是"说得好听的话，空口的应酬话"，when it gets in their way 是状语从句，when 前面和后面的 it 均指网络安全问题。
译文：许多人大肆空谈安全，但是当安全问题妨碍到他们时，又不想麻烦。

Exercises

Ⅰ. Put "true" or "false" in the brackets for the following statements according to the passage.

1. ( ) In this article, "cybermarket" means a market that sells all kinds of goods.
2. ( ) In Internet communication, we show concern about secrecy and authentication.
3. ( ) The two kinds of encryption techniques used in the Internet are symmetric key encryption and asymmetric key encryption.

4. ( ) Checking the person's identification is one of the CA's tasks.

5. ( ) Internet delivers information fast and it is not restricted by time and place.

6. ( ) In person-to-person community, security is based on physical cues.

7. ( ) Encryption is the only way to solve the data security problem.

8. ( ) The advantage of using symmetric key encryption lies in its fast encryption and decryption processes.

9. ( ) The private key is available to the public.

10. ( ) The lack of strong authentication has inhibited the development of electronic commerce.

II. Fill in the blanks according to the passage.

1. _____ people use Internet, _____ concerns about Internet security.

2. Many people pay great amounts of _____ to security, but do not want to be bothered with it when it gets in their way.

3. When a person wants to get a digital certificate, he generates his own key pair, gives the _____ as well as some proof of his identification to the Certificate Authority.

4. There are two kinds of encryption techniques — _____ key encryption and _____ key encryption.

5. Digital certificate is an identity card _____ in the computer society.

6. The lack of strong authentication has _____ the development of electronic commerce.

7. Strong _____ is then, a key requirement if the Internet is to be used for electronic commerce.

8. Users who find security policies and systems too _____ will find ways around them.

9. To name but a few, we use our _____ to authenticate ourselves.

10. For symmetric key encryption, both parties should have a _____ about a secret encryption key.

III. Translate the following words and expressions into Chinese.

1. up-to-date   6. identity card
2. symmetric key encryption 7. digital certificate
3. asymmetric key encryption 8. Certificate Authority
4. vice versa   9. a trustworthy certificate
5. as well as   10. one-time

6.3.1 Reading Material

Internet Firewall Concept

A packet filter is often used to protect an organization's computers and networks from unwanted Internet traffic. The filter is placed in the router that connects the organization to the rest of the Internet.

A packet filter configured to protect an organization against traffic from the rest of the

Internet is called an Internet firewall; the term is derived from the fireproof physical boundary placed between two structures to prevent fire from moving between them. Like a conventional firewall, an Internet firewall is designed to keep problems in the Internet from spreading to an organization's computers.

Firewalls are the most important security tool used to handle network connections between two organizations that do not trust each other. By placing a firewall on each external network connection, an organization can define a secure perimeter that prevents outsiders from interfering with the organization's computers. In particular, by limiting access to a small set of computers, a firewall can prevent outsiders from probing all computers in an organization or flooding the organization's network with unwanted traffic.

A firewall can lower the cost of providing security. Without a firewall to prevent access, outsiders can send packets to arbitrary computers in an organization. Consequently, to provide security, an organization must make all of its computer secure. With a firewall, however, a manager can restrict incoming packets to a small set of computers. In the extreme case, the set can contain a single computer. Although computers in the set must be secure, other computers in the organization do not need to be. Thus, an organization can save money because it is less expensive to install a firewall than to make all computer systems secure.

Words

boundary	*n.*	边界，分界线
consequently	*adv.*	结果；因此
external	*adj.*	外部的，客观的
	n.	外部，外面
fireproof	*adj.*	耐火的，防火的
firewall	*n.*	防火墙
interfere	*vi.*	干涉，干预，妨碍，打扰
packet	*n.*	信息包
	v.	包装
perimeter	*n.*	周；周围；周界筑有防御工事的地带或边界
probe	*n.*	探针，探测器
	vt.	探查，查明

6.3.2 正文参考译文

网络安全

近几年来，Internet 使人们的生活发生了很多改变。人们使用 e-mail、通过 IP 电话和朋友交谈，从网上获取最新信息，在网络市场购物。与传统通信渠道相比，Internet 有许多优势：成本低、效率高、信息传送速度快、并且不受时间和地点的限制。使用 Internet 的人越多，对 Internet 安全的关注就越多。

人与人之间直接交流，安全取决于物理线索。略举几例：人们用签名来表明自己的身份；人们把信函密封起来，防止他人窥视和更改；人们接受商店里有公章的收条来证明已经付款；人们从可靠的地方获取信息。不过对 Internet 安全而言，就没有这样的物理提示。对 Internet 通信，人们关心两个方面，第一是保密——如何确信数据在传输过程中没有人阅读过？第二是认证——如何确信某个人(或计算机)所声称的身份。

　　解决数据安全问题的途径是加密。在现实生活中，如果汤姆想和玛丽密谈，他可以找一间没人的房子和玛丽悄悄地交谈，或者他用只有他们两人明白的密码交谈。在 Internet，人们用第二种方法——加密来传输数据。加密技术有两种——对称密钥加密和非对称密钥加密。

　　对对称密钥加密来说，当事人双方要有一致的密钥。当 A 要给 B 发送消息时，A 用密钥将消息加密。B 收到加密的消息后，用相同的(或最初的)密钥将消息解密。对称密钥加密的优点在于它的加密和解密速度快(与相同安全标准下的非对称密钥加密术相比)。它的缺点是：第一，在发送秘密消息之前，当事双方必须安全地交换密钥；第二，对不同的当事人，人们必须使用不同的密钥。例如，如果 A 和 B、C、D 及 E 通信，A 必须用四种不同的密钥。否则，B 将知道 A 和 C 以及 A 和 D 在谈论什么。要找到安全交换密钥的方式很困难，所以，对称密钥加密的缺点使它不适合用于 Internet。

　　对非对称密钥加密，当事各方都有一对密钥：公钥和私人密钥。公钥可自由使用，但只有密钥持有者拥有私人密钥。用公钥加密的消息只能用相应的私人密钥解密，反之亦然。当 A 给 B 发送消息时，A 首先得到 B 的公钥将消息加密，然后发送给 B。B 收到消息后，用他的私人密钥将消息解密。这种加密术的优点是人们可以自由获得公钥，因此可被从交换密钥问题中解脱出来。它的缺点是加密和解密速度慢。非对称密钥加密在数据传输上似乎是安全的，但依然存在认证问题，例如，当 A 给 B 发送消息时，A 从互联网上得到 B 的公钥，但 A 怎样才能知道他获得的公钥确实属于 B？这个问题由数字证书来解决。

　　数字证书相当于电脑世界的身份证。当一个人想获得数字证书时，他生成自己的一对密钥，把公钥和其他的鉴定证据送达证书授权机构，证书授权机构将核实这个人的资料，来确定申请人的身份。如果申请人确如自己所声称的，证书授权机构将授予带有申请人姓名、电子邮件地址和申请人公钥的数字证书，并且该数字证书由证书授权机构用其私有密钥进行数字签名。当 A 要给 B 发送消息时，A 必须得到 B 的数字证书，而非 B 的公钥。A 首先核实带有证书授权机构公钥的签名，以确定是否为可信赖的证书。然后，A 从证书上获得 B 的公钥，并利用公钥将消息加密后送给 B。

　　认证是日常生活中的重要部分。缺少强有力的认证制约了电子商务的发展，这就使得写在纸上的合同、法律文件和官方信函仍是必要的。如果互联网用于电子商务，强有力的认证是一个关键要求。强有力的认证通常建立在现代版的一次性密码本技术上。例如，令牌用来代替昔日的一次性密码本，而且储存在智能卡或磁盘上。

　　许多人大肆空谈安全，但是当安全问题妨碍到他们时，又不想麻烦。建立一个无需用户时刻想起周围的安全保障系统的系统和网络就非常重要。如果用户觉得安全政策和系统过于约束，他们就会设法绕过安全政策和系统。安全是每个人的事情，只有通过人人协作、采用明智的政策并坚持实践，网络安全才能实现。

6.3.3 阅读材料参考译文

Internet 防火墙的概念

数据包过滤器通常用来保护机构的计算机和网络，使之过滤掉不想要的互联网流量。该过滤器放置在连接该机构与互联网的路由器中。

配置用来保护一个机构免受其他互联网流量影响的数据包过滤器被称为 Internet 防火墙，"防火墙"一词源自于放置在两个机构之间，以防止火灾在它们之间蔓延的防火物理边界。像传统的防火墙，Internet 防火墙的目的是防止互联网中的问题扩散到一个机构内的计算机。

作为最重要的安全工具，防火墙用来处理两个互不信任的组织之间的网络连接。通过在每个外部网络连接的地方放置一个防火墙，机构可以定义一个安全的边界，防止外界干扰本机构的计算机。通过使访问限于一小组计算机，防火墙尤其可以防止外界探测所有计算机或者阻止不想要的流量涌向该组织的网络。

防火墙能够降低提供安全保障的成本。如果没有防火墙来防止访问，外界可以向该机构内任意计算机发送数据包。因此，为了提供安全保障，必须保证所有计算机的安全。管理员使用防火墙可以使传入的数据包只进入一小组计算机。在极端情况下，小组只有一台计算机。尽管组中的计算机必须受保护，但是该机构中的其他计算机不需要。这样，一个机构可以节约开支，因为安装一个防火墙比对所有计算机系统进行安全保护花费少。

6.4 Secure Networks and Policies

What is a secure network? Can the Internet be made secure? [1] Although the concept of a secure network is appealing to most users, networks cannot be classified simply as secure or not secure because the term is not absolute—each group defines the level of access that is permitted or denied. For example, some organizations store data that is valuable. Such organizations define a secure network to be a system that prevents outsiders from accessing the organization's computers. Other organizations need to make information available to outsiders, but prohibit outsiders from changing the data. Such organizations may define a secure network as one that allows arbitrary access to data, but includes mechanisms that prevent unauthorized changes. Finally, many large organizations need a complex definition of security that allows access to selected data or services the organization chooses to make public, while preventing access or modification of sensitive data and services that are kept private.

[2]Because no absolute definition of information secure exists, the first step an organization must take to achieve a secure system is to define the organization's security policy. The policy does not specify how to achieve protection. Instead, it states clearly and unambiguously the items that are to be protected.

Defining an information security policy is complex. The primary complexity arises because an information security policy cannot be separated from the security policy for computer systems attached to the network. In particular, defining a policy for data that traverses a network does not guarantee that data will be secure. Information security cannot prevent unauthorized users who have accounts on the computer from obtaining a copy of the data. The policy must hold for the data stored on disk, data communicated over a telephone line with a dialup modem, information printed on paper, data transported on portable media such as a floppy disk, and data communicated over a computer network.

Defining a security policy is also complicated because each organization must decide which aspects of protection are most important, and often must compromise between security and ease of use. For example, an organization can consider:

- Data Integrity;
- Data Availability;
- Data Confidentiality and Privacy.

Words

accountability	n.	责任，可计算性
archive	v.	存档
	n.	档案文件
authorization	n.	授权，特许
incur	v.	招致
liability	n.	责任，债务，负债
traverse	n.	横贯，横断
	v.	横过，穿过，经过

Phrases

appeal to	呼吁，要求，诉诸，上诉，有吸引力
arise from	起于，由……而引起，由……而产生
be separated from	和……分离开，和……分散
data integrity	数据完整性
data availability	数据有效性
data confidentiality	数据机密性
focus on	集中
integrity mechanisms	完整性机制
prevent…from	阻止，防止
prohibit…from	禁止，阻止

Notes

[1] 例句：Although the concept of a secure network is appealing to most users, networks cannot be classified simply as secure or not secure because the term is not absolute—each group defines the level of access that is permitted or denied.

Chapter 6　Information Security

　　分析：这是一个复合句。主句为 networks cannot be classified simply as secure or not secure, be classified as 意思是"被分类为"。

　　译文：尽管安全网络的概念吸引着大多数用户，但我们不能把网络简单地分为安全的或不安全的，因为安全这个术语不是绝对的——每个团体规定允许访问或拒绝访问的标准。

[2] 例句：Because no absolute definition of information secure exists, the first step an organization must take to achieve a secure system is to define the organization's security policy.

　　分析：本句为复合句。主句中的 an organization must take to achieve a secure system 为定语从句，修饰主语 the first step，to define the organization's security policy 为动词不定式作表语。

　　译文：因为不存在信息安全的绝对定义，一个机构要获得安全系统的第一步就是定义机构的安全政策。

Exercises

Ⅰ. Put "true" or "false" in the brackets for the following statements according to the passage.

1. (　) Networks can be classified as secure or not secure.
2. (　) Security policy specifies how to achieve protection.
3. (　) Defining a policy for data that traverses a network does not guarantee that data will be secure.
4. (　) A security policy cannot be defined unless an organization understands the value of its information.
5. (　) Many large organizations need a complex definition of security that allows access to selected data or services the organization chooses to make public.
6. (　) An absolute definition of information secure exists.
7. (　) Information security must prevent unauthorized users who have accounts on the computer from obtaining a copy of the data.
8. (　) Each group defines the level of access that is permitted or denied.
9. (　) Other organizations need to make information available to outsiders, but prohibit outsiders from changing the data.
10. (　) Defining a security policy isn't complicated.

Ⅱ. Fill in the blanks according to the passage.

1. _____ refers to protection from change: is the data that arrives at a receiver exactly the same as the data that was sent?

2. _____ refers to protection against disruption of service: does data remain accessible for legitimate uses?

3. _____ refer to protection against snooping or wiretapping: is data protected against unauthorized access?

4. Defining an information Security Policy is _____.

5. Networks cannot be _____ simply as secure or not secure because the term is not absolute—each group defines the level of access that is permitted or denied.

6. The concept of a secure network is _____ to most users.

7. Such organizations define a secure network to be a system that prevents outsiders from _____ the organization's computers.

8. Many large organizations need a complex _____ of security.

9. The first step an organization must take to _____ a secure system is to define the organization's security policy.

10. Defining a policy for data that _____ a network does not guarantee that data will be secure.

III. Translate the following words into Chinese.

1. appeal to
2. data integrity
3. data availability
4. data confidentiality
5. integrity mechanisms
6. prohibit…from
7. be responsible for
8. be analogous to
9. result from
10. information security

6.4.1 Reading Material

Computer Security

The techniques developed to protect single computers and network-linked computer systems from accidental or intentional harm are called computer security. Such harm includes destruction of computer hardware and software, physical loss of data, and the deliberate invasion of databases by unauthorized individuals.

Data may be protected by such basic methods as locking up terminals and replicating data in other storage facilities. Most sophisticated methods include limiting data access by requiring the user to have an encoded card or to supply an identification number or password. Such procedures can apply to the computer-data system as whole or may be pinpointed for particular information banks or programs. Data are frequently ranked in computer files according to degree of confidentiality.

Operating systems and programs may also incorporate built-in safeguards, and data may be encoded in various ways to prevent unauthorized persons from interpreting or even copying the material. The encoding system most widely used in the United States is the Data Encryption Standard (DES), designed by IBM and approved for use by the National Institute of Standards and Technology in 1976. DES involves a number of basic encrypting procedures that are then repeated several times. Very large-scale computer systems, for example, the U.S. military's Advanced Research Project Agency Network (ARPANET), may be broken up into smaller subsystems for security purposes, but smaller system in government and industry are more prone to system-wide invasions. At the level of personal computers, security possibilities are fairly minimal.

Most invasions of computer systems are for international or corporate spying or sabotage, but computer hackers may take the penetration of protected databanks as a challenge, often with no object in mind other than accomplishing a technological feat. Of growing concern is the deliberate implantation in computer programs of worms or viruses that, if undetected, may progressively destroy databases and other software. Such infected programs have appeared in the electronic bulletin boards available to computer users. Other viruses have been incorporated into computer software sold commercially. No real protection is available against such bugs except the vigilance of manufacturer and user.

Words

accomplish	vt.	完成，达到，实现
break up	v.	打碎，分裂，分解
bulletin	n.	公告，报告
confidentiality	n.	机密性
deliberate	adj.	深思熟虑的，故意的，预有准备的
destruction	n.	破坏，毁灭
facility	n.	容易，灵巧，熟练，设备，工具
feat	n.	技艺，本领
implantation	n.	培植，灌输，建立
intentional	adj.	有意的，故意的
invasion	n.	入侵
penetration	n.	穿过，渗透，突破，入侵
pinpoint	adj.	针尖的，极微小的，精确定点(位)的，细致的
progressively	adv.	日益增多地
prone	adj.	倾向于
sabotage	n.	怠工，破坏
vigilance	n.	警惕，警觉

Phrases

computer security	计算机安全
computer virus	计算机病毒

Abbreviations

DES(Data Encryption Standard)	数据加密标准
IBM (International Business Machines)	(美国)商用机器公司

6.4.2 正文参考译文

网络安全和政策

什么是安全网络？Internet 是安全的吗？尽管安全网络的概念吸引着大多数用户，但我们不能把网络简单地分为安全的或不安全的，因为安全这个术语不是绝对的——每个团体都有规定的允许访问或拒绝访问的标准。例如，一些机构存储着重要数据，这样的机构可能把安全网络定义为系统能够防止外界对本机构计算机的访问。其他一些机构需要向外界提供有效的信息，但要禁止外界对数据的更改。这样的机构可能把安全网络定义为允许任意访问数据、但是要有能够防止非法更改数据的机制。最终，许多大型机构需要一个对安全的复杂定义，这种安全允许访问本机构对外公开的部分数据和服务，同时又禁止对其处于保密状态的敏感数据和服务进行访问或修改。

因为不存在信息安全的绝对定义，一个机构要获得安全系统的第一步就是定义自己机构的安全政策。政策不规定如何去实现保护，而是要清楚明白地表明哪些项目需要得到保护。

定义信息安全政策非常复杂。首要的复杂性在于信息安全政策与网络中的计算机系统的安全政策密不可分。尤其为横贯于网络中的数据定义安全政策时无法保证其数据的安全性。信息安全不能禁止在计算机上拥有账户的非法使用者获得数据的副本。所以安全政策必须针对存储在磁盘上的数据、通过拨号调制解调器用电话线传送的数据、书面形式打印出来的信息、通过便携的媒介如软盘传送的数据以及通过计算机网络传送的数据等。

定义安全政策也是复杂的，因为每个机构必须决定哪些方面的保护是最重要的，而且常常要在安全和易于使用之间进行折中处理。例如，某个机构可以考虑以下几个方面：
- 数据完整性；
- 数据可用性；
- 数据机密性和保密。

6.4.3 阅读材料参考译文

计算机安全

为保护单一的计算机和联网的计算机系统，以防发生意外或故意损害而开发的技术称为计算机安全。对计算机的伤害包括破坏计算机硬件和软件、丢失数据、以及未经授权的个人蓄意入侵数据库。

我们可以用锁定终端和在其他存储设施复制数据等基本方法来保护数据。最先进的方法包括限制数据存取，要求用户有一个编码卡或提供身份证号码或密码。这种程序可用于整个计算机数据系统或针对特定的资料库或程序。计算机文件中的数据往往根据机密程度来排列。

操作系统和程序一般包含内置的安全保障，数据可能以各种方式编码以防止未经授权的人改写甚至抄袭。在美国最广泛使用的编码系统是数据加密标准(DES)。DES 由 IBM 设计，1976 年由美国国家标准与技术研究院核准使用。DES 算法包含几个重复数次的基本加密过程。为了安全起见，超大型计算机系统，例如美国国防部高级研究计划局建立的计算机网(ARPANet)，可细分成更小的子系统。但在政府和工业界，规模较小的系统更容易遭到系统入侵。个人电脑的安全性微乎其微。

大多数计算机系统入侵是国际或企业的间谍行为或蓄意破坏，但电脑黑客却把侵入受保护的数据库当作挑战，除了完成一项技术壮举外没有其他目的。目前日益引起关注的问题是故意植入计算机程序的病毒或蠕虫，如果未被发现，这些病毒或蠕虫可能会逐步销毁数据库和其他软件，这种感染的程序已经出现在提供给计算机用户使用的电子公告板中。其他病毒已经进入商业销售的计算机软件。针对这种问题，除了制造商和用户保持高度的警惕外，没有真正的病毒保护措施可用。

Chapter 7   Image Processing

7.1   Concepts of Graphic and Image

Image or Graphic? Technically, neither.[1]If you really want to be strict, computer pictures are files, the same way WORD documents or solitaire games are files. They're all a bunch of ones and zeros all in a row. But we do have to communicate with one another, so let's decide.

Image. We'll use "image". That seems to cover a wide enough topic range.

"Graphic" is more of an adjective, as in "graphic format". we denote images on the Internet by their graphic format. GIF is not the name of the image. GIF is the compression factor used to create the raster format set up by CompuServe.

So, they're all images unless you're talking about something specific.

The images produced in Drawing programs (CorelDraw, Illustrator, Freehand, Designer etc) are called vectorised graphics. [2]That is, all of the objects shown on the computer monitor are representations of points and their relationship to each other on the work area, each of which is stored in the computer as simple values and mathematical equations depicting: the relationship between each point and the next point referenced to it, and the position (vector) of each point referenced to a starting corner of the work area.

Bitmap pictures are stored as a vertical and horizontal array of Pixels and stored information represents the colour of each of these pixels. The resolution of a bitmap picture describes how many of these pixels exist over a set distance, usually horizontally: i.e. pixels per inch or pixels per centimetre. An unaltered bitmap picture of 300 pixels/inch enlarged by 1000% will therefore still have the same number of pixels across the actual picture area but each represented pixel will cover a larger area.

[3]At such an enlargement, the picture would be of little use for reproduction unless viewed from quite a long distance.

Bitmap or Photo-retouching programs are correctly called PAINTING PROGRAMS.

Vectorised drawings on the other hand can be enlarged as much as desired. because, although the above mentioned points on a drawing would be further apart, the relationship of any described line between the points would always be the same. A single company logo file produced in a Drawing program could be used for a business card, any brochure or poster, or plotting out to a Screen Print stencil 3 metres (9 feet) wide, where as bitmap files would have to be created for every size used if practicable.

What is raster, vector, metafile, PDL, VRML, and so forth?

These terms are used to classify the type of data a graphics file contains.

Raster files (also called bitmapped files) contain graphics information described as pixels, such as photographic images. Vector files contain data described as mathematical equations and are typically used to store line art and CAD information. Metafiles are formats that may contain either raster or vector graphics data. Page Description Languages (PDL) are used to describe the layout of a printed page of graphics and text.

Animation formats are usually collections of raster data that is displayed in a sequence. Multi-dimensional object formats store graphics data as a collection of objects (data and the code that manipulates it) that may be rendered (displayed) in a variety of perspectives. Virtual Reality Modeling Language (VRML) is a 3D, object-oriented language used for describing "virtual worlds" networked via the Internet and hyperlinked within the World Wide Web. Multimedia file formats are capable of storing any of the previously mentioned types of data, including sound and video information.

Words

brochure	*n.*	小册子
bunch	*n.*	串，束
	v.	捆成一束
depict	*v.*	描述，描写
enlarge	*v.*	扩大，放大
metafile	*n.*	元文件，图元文件
perspective	*n.*	透视图，远景，观点，观察
pixel	*n.*	(显示器或电视机图像的)像素
poster	*n.*	海报，宣传画
practicable	*adj.*	能实行的，可行的
raster	*n.*	[物]光栅
render	*v.*	呈递，归还，着色，给予补偿
representation	*n.*	表示法，表现，代表
solitaire	*n.*	单人纸牌游戏
stencil	*n.*	模版
vector	*n.*	[数]向量，矢量

Phrases

compression factor	压缩因子
CompuServe	美国最大的在线信息服务机构之一

Abbreviations

CAD(Computer Aided Design)	计算机辅助设计
PDL(Page Description Languages)	页面描述语言
VRML(Virtual Reality Modeling Language)	虚拟现实建模语言

Notes

[1] 例句：If you really want to be strict, computer pictures are files, the same way WORD documents or solitaire games are files.

分析：本句是复合句，主句中 the same way WORD documents or solitaire games are files 为方式状语，the same way 前省去了 in。

译文：从严格意义上来说，计算机图像是文件，就像 Word 文档或纸牌游戏是文件一样。

[2] 例句：That is, all of the objects shown on the computer monitor are representations of points and their relationship to each other on the work area, each of which is stored in the computer as simple values and mathematical equations depicting: the relationship between each point and the next point referenced to it, and the position (vector) of each point referenced to a starting corner of the work area.

分析：本句是复合句。that is 为插入语，意思为 "即；也就是说"。each of which is stored in the computer as simple values and mathematical equations… 为 "名词+介词+which" 开始的定语从句，修饰 representations。

译文：也就是说，在计算机监视器显示的所有对象都是工作区域内的点和它们之间相互关系的表示，每一点都作为简单值或数学公式存储在计算机里，其中数学公式用来描述每个点和参照该点的下一个点之间的关系和参照工作区域开始角的每个点的位置(矢量)。

[3] 例句：At such an enlargement, the picture would be of little use for reproduction unless viewed from quite a long distance.

分析：本句是复合句。句中 unless viewed from quite a long distance 为 unless 引导的真实条件状语从句，unless 通常相当于 if not，unless 后省去了 it is。

译文：在这样一个放大图片里，图像无法用于再现，除非从相当远的距离外观看。

Exercises

Ⅰ. Put "true" or "false" in the brackets for the following statements according to the passage.

1. (　) Computer pictures are files that are all a bunch of ones and zeros all in a row.
2. (　) GIF is the name of the image.
3. (　) Bitmap pictures are stored as a vertical and horizontal array of Pixels.
4. (　) The stored information represents the colour of each of the pixels.
5. (　) Vectorised drawings can be enlarged as much as desired.
6. (　) Bitmap pictures can not be enlarged as much as desired.
7. (　) Raster files contain graphics information described as pixels.
8. (　) Vector files contain data described as mathematical equations.
9. (　) Multimedia file formats are capable of storing sound and video information.
10. (　) Animation formats are usually collections of raster data that is displayed in a sequence.

II. Fill in the blanks according to the passage.

1. Computer pictures are _____, the same way WORD documents or solitaire games are _____.

2. The _____ of a bitmap picture describes how many of these pixels exist over a set distance, usually _____.

3. Raster, _____, metafile, PDL, and VRML are used to _____ the type of data a graphics file contains.

4. Raster files (also called bitmapped files) contain graphics information described as _____.

5. Photo-retouching programs are correctly called _____.

6. Vector files contain data described as _____ and are typically used to store line art and CAD _____.

7. _____ file formats are capable of storing sound and video information.

8. Page Description Languages (PDL) are used to describe the _____ of a printed page of graphics and _____.

9. Files are all a bunch of _____ and _____ all in a row.

10. GIF is the _____ factors used to create the raster format set up by _____.

III. Translate the following words and phrases into Chinese.

1. row   6. mathematical equation
2. image   7. monitor
3. produce   8. layout
4. array   9. format
5. enlargement   10. store

7.1.1 Reading Material

Some major graphic file formats

There are hundreds of image file types. The PNG, JPEG, and GIF formats are most often used to display images on the Internet.

JPEG (Joint Photographic Experts Group) files are (in most cases) a lossy format; the DOS filename extension is JPG (other operating systems may use JPEG). Nearly every digital camera can save images in the JPEG format, which supports 8 bits per color (red, green, blue) for a 24-bit total, producing relatively small files. When not too great, the compression does not noticeably detract from the image's quality, but JPEG files suffer generational degradation when repeatedly edited and saved. Photographic images may be better stored in a lossless non-JPEG format if they will be re-edited, or if small "artifacts" (blemishes caused by the JPEG's compression algorithm) are unacceptable. The JPEG format also is used as the image compression algorithm in many Adobe PDF files.

The PNG (Portable Network Graphics) file format was created as the free, open-source successor to the GIF. The PNG file format supports true color (16 million colors) while the GIF supports only 256 colors. The PNG file excels when the image has large, uniformly colored areas.

Chapter 7 Image Processing

The lossless PNG format is best suited for editing pictures, and the lossy formats, like JPG, are best for the final distribution of photographic images, because JPG files are smaller than PNG files. Many older browsers currently do not support the PNG file format, however, with Mozilla Firefox or Internet Explorer 7, all contemporary web browsers now support all common uses of the PNG format. PNG, an extensible file format for the lossless, portable, well-compressed storage of raster images. PNG provides a patent-free replacement for GIF and can also replace many common uses of TIFF. Indexed-color, grayscale, and truecolor images are supported, plus an optional alpha channel. PNG is designed to work well in online viewing applications, such as the World Wide Web. PNG is robust, providing both full file integrity checking and simple detection of common transmission errors. Also, PNG can store gamma and chromaticity data for improved color matching on heterogeneous platforms.

GIF (Graphics Interchange Format) is limited to an 8-bit palette, or 256 colors. This makes the GIF format suitable for storing graphics with relatively few colors such as simple diagrams, shapes, logos and cartoon style images. The GIF format supports animation and is still widely used to provide image animation effects. It also uses a lossless compression that is more effective when large areas have a single color, and ineffective for detailed images or dithered images.

Words

artifacts	*n.*	史前古器物
blemish	*n.*	污点，缺点，瑕疵
	v.	弄脏，玷污，损害
chromaticity	*n.*	染色性，色品(度)
contemporary	*adj.*	同时代的，同年龄的，当代的
degradation	*n.*	降级，降格，退化
detract	*v.*	转移，诽谤，贬低，毁损
dithered	*adj.*	[计] 颤动的
extensible	*adj.*	可扩展的，可延伸的
extension	*n.*	延长，扩充，范围，扩展名
generational	*adj.*	世代的
heterogeneous	*adj.*	不同的，异类的
logo	*n.*	标识语
lossless	*adj.*	无损的
lossy	*adj.*	[电]有损耗的，致损耗的
palette	*n.*	调色板，颜料
photographic	*adj.*	摄影的，详细记录的
portable	*adj.*	轻便的，便携式的，便于携带的
robust	*adj.*	强壮的，坚固的，耐用的
successor	*n.*	继承者，接任者，后续的事物

7.1.2 正文参考译文

graphic 和 image 的概念

计算机图像是 graphic 还是 image？从技术上说都不是。严格地说，计算机图画是文件，就像 Word 文档或纸牌游戏是文件一样。它们都是由很多 1 和 0 排成的数据。但是由于我们需要互相交流，所以要确定名称。

"image"。我们要使用"image"的概念。这一概念似乎覆盖了非常广泛的话题范围。(graphic 经常翻译成图形、绘图，image 翻译成图像、影像。编者注)

"graphic"更多的情况是作形容词用，例如"图形格式"。我们用图形格式来表示 Internet 中的图像。GIF 不是图像的名字，而是用来生成通过 CompuServe 建立的光栅格式的压缩因子。

所以除非被用来讨论专用的事物，它们都是图像。

绘图程序(CorelDraw、Illustrator、Freehand、Designer 等)生成的图像都被称为矢量化图形。也就是说，在计算机显视器显示的所有对象都是工作区域内的点和它们之间相互关系的表示，每一点都作为简单值或数学公式存储在计算机里，其中数学公式用来描述每个点和参照该点的下一个点之间的关系和参照工作区域开始角的每个点的位置(矢量)。

位图图像存储为垂直和水平排列的像素，存储的像素信息表示每个像素的颜色。位图图像的清晰度描述在一个特定距离(通常指水平距离)内有多少像素：即像素数/英寸或像素数/厘米。一个未改变的 300 像素/英寸的位图图像放大 10 倍后在实际的图像范围内将会有相同的像素数目，但每一个表示出的像素将会覆盖更大的范围。

在这样一个放大的图片里，图像无法用于再现，除非从相当远的距离外观看。

位图或照片润色程序的正确叫法为画图程序。

另一方面，矢量化的图形可以按要求尽量放大。因为尽管上面提到的图形上的点可以相隔得更远，但点之间任一描述线之间的关系是相同的。也就是说，画图程序产生的公司标语文件可以用作名片、小册子或海报，或绘制到一个 3 米(9 英尺)宽的打印模板上，如果可行，需要为每个可用尺寸建立位图文件。

光栅、矢量、图元文件、PDL、VRML 等是什么呢？

这些术语用来划分一个图像文件含有的数据种类。

光栅文件也被称为位图文件，包含以像素描述的图像信息，例如摄影图像。矢量文件含有被描述为数学等式的数据，并通常用来存储矢量图或 CAD 信息。图元文件包含有光栅或矢量图数据。页面描述语言(PDL)被用来描述图形或文本打印页面的版面设置。

动画格式通常是以序列显示的光栅数据集。多维对象格式将图像数据作为在各种视角下表示(展示)的对象集(数据及操作数据的代码)存储起来。VRML(虚拟现实建模语言)是用来描述通过 Internet 联网或全球网链接的"虚拟世界"的三维面向对象的语言。多媒体文件格式可以存储前面提到的任一种数据，包括声音和视频信息。

7.1.3 阅读材料参考译文

几种主要的图像文件格式

图像文件的类型有上百种，常用于在 Internet 显示图像的文件格式有 PNG、JPEG 和 GIF。

JPEG(联合图像专家组)文件(在很多情况下)为有损压缩，DOS 下文件扩展名是 JPG(其他操作系统下可能使用 JPEG)。几乎所有的数码相机都可以用 JPEG 格式保存图像，JPEG 支持 24 位色彩，每个颜色(红色、绿色、蓝色)8 位，产生相对小的文件。在文件不太大的情况下，压缩不会明显降低图像质量，但 JPEG 文件在多次编辑与保存时会逐次降低质量。为了再编辑，或者如果无法接受由 JPEG 压缩算法引起的小的瑕疵，摄影图片最好保存为无损图像质量的非 JPEG 格式。在许多 Adobe PDF 文件中，也使用 JPEG 格式作为图像压缩算法。

PNG(可移植的网络图像)文件格式是作为免费开放源代码的 GIF 后续而开发的。PNG 文件格式支持真彩色(16 000 000 种颜色)，而 GIF 格式只支持 256 色。在图像具有大幅均匀色彩区域的情况下，PNG 文件占优势。无损的 PNG 格式最适用于编辑图画，像 JPG 这样的有损格式最适用于照片图像的最后发行，这是因为 JPG 文件比 PNG 文件小得多。很多老的浏览器不支持 PNG 格式，但是，有了 Mozilla Firefox 或 Internet Explorer 7，所有当代的网络浏览器都支持 PNG 格式的一般应用。PNG 是一个可扩充的文件格式，用于光栅图像的无损、可移植、良好的压缩存储。PNG 提供一个非专利的 GIF 格式的替代，并且可以替代许多 TIFF 的一般应用。支持色彩索引、灰度、真彩色以及可选的 alpha 通道，PNG 设计用于在线浏览应用，如 World Wide Web。PNG 是耐用的，提供完整的文件完整性检测和一般传输错误的简单检测。并且 PNG 可以存储 gamma 值和色度数据，以在不同平台上改进色彩匹配。

GIF(图形交换格式)局限于 8 位调色板或者 256 色。这使得 GIF 格式适用于存储颜色相对少的图形如简单图表、形状、标识和卡通风格的图像。GIF 格式支持动画并且广泛用于提供动画效果。GIF 格式也使用无损压缩，这种压缩在单色画面较大的情况下更有效，而对细节多或颤动的图像压缩效率则很低。

7.2 Introduction to Digital Image Processing

An image is digitized to convert it to a form that can be stored in a computer's memory or on some form of storage media such as a hard disk or CD-ROM. This digitization procedure can be done by a scanner, or by a video camera connected to a frame grabber board in a computer. Once an image has been digitized, it can be operated upon by various image processing operations.

Image processing operations can be roughly divided into three major categories, Image Compression, Image Enhancement and Restoration, and Measurement Extraction. Image compression is familiar to most people. [1]It involves reducing the amount of memory needed to store a digital image.

Image defects which could be caused by the digitization process or by faults in the imaging set-up (for example, bad lighting) can be corrected using Image Enhancement techniques. Once the image is in good condition, the Measurement Extraction operations can be used to obtain useful information from the image.

Some examples of Image Enhancement and Measurement Extraction are given below. The examples shown all operate on 256 grey-scale images. This means that each pixel in the image is stored as a number between 0 to 255, where 0 represents a black pixel, 255 represents a white pixel and values in-between represent shades of grey. These operations can be extended to operate on colour images.

The examples below represent only a few of the many techniques available for operating on images. Details about the inner workings of the operations have not been given.

Image Enhancement and Restoration

The image at the left of Figure 7.1 has been corrupted by noise during the digitization process. The "clean" image at the right of Figure 7.1 was obtained by applying a median filter to the image.

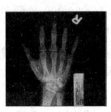 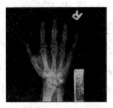

An image with poor contrast, such as the one at the left of Figure 7.2, can be improved by adjusting the image histogram to produce the image shown at the right of Figure 7.2.

Figure 7.1   Application of the median

Figure 7.2   Adjusting the image histogram to improve image contrast

The image at the top left of Figure 7.3 has a corrugated effect due to a fault in the acquisition process. This can be removed by doing a 2-dimensional Fast-Fourier Transform on the image (top right of Figure 7.3), removing the bright spots (bottom left of Figure 7.3), and finally doing an inverse Fast Fourier Transform to return to the original image without the corrugated background (bottom right of Figure 7.3).

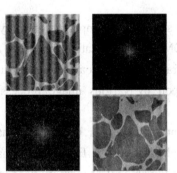

An image which has been captured in poor lighting conditions and shows a continuous change in the background brightness across the image (top left of Figure 7.4) can be corrected using the following procedure. First remove the foreground objects by applying a 25 by 25

Figure 7.3   Application of the 2-dimensional Fast Fourier Transform

greyscale dilation operation (top right of Figure 7.4). Then subtract the original image from the background image (bottom left of Figure 7.4). Finally invert the colors and improve the contrast by adjusting the image histogram (bottom right of Figure 7.4).

Chapter 7 Image Processing

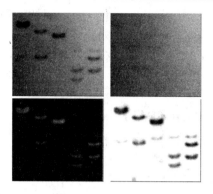

Figure 7.4 Correcting for a background gradient

Image Measurement Extraction

[2]The example below demonstrates how one could go about extracting measurements from an image. The image at the top left of Figure 7.5 shows some objects. The aim is to extract information about the distribution of the sizes (visible areas) of the objects. The first step involves segmenting the image to separate the objects of interest from the background. This usually involves thresholding the image, which is done by setting the values of pixels above a certain threshold value to white, and all the others to black (top right of Figure 7.5). Because the objects touch, thresholding at a level which includes the full surface of all the objects does not show separate objects. This problem is solved by performing a watershed separation on the image (lower left of Figure 7.5). The image at the lower right of Figure 7.5 shows the result of performing a logical AND of the two images at the left of Figure 7.5. This shows the effect that the watershed separation has on touching objects in the original image. Finally, some measurements can be extracted from the image.

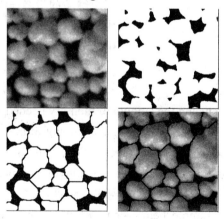

Figure 7.5 Thresholding an image and applying a Watershed Separation Filter

Words

2-dimensional	*adj.*	二维的
acquisition	*n.*	获得，获得物

contrast	v.	使与……对比，和……形成对照
	n.	对比
convert	v.	使转变，转换……
corrugate	v.	弄皱，(使)成波状
corrupt	adj.	被破坏的
	v.	使恶化，堕落
digitize	v.	将资料数字化
dilation	n.	膨胀，扩大
enhancement	n.	增进，增加
extraction	n.	提取
histogram	n.	柱状图
improved	adj.	改良的
invert	adj.	转化的
	v.	使颠倒
	n.	颠倒的事物
measurement	n.	测量
restoration	n.	恢复，修补
roughly	adv.	概略地，粗糙地
threshold	n.	开始，开端，极限
watershed	n.	分水岭

Prases

Fast-Fourier Transform	快速傅里叶变换
frame grabber board	帧中继访问设备(=FRAD)
inverse Fast Fourier Transform	快速傅里叶反变换
median filter	中值滤波器

Notes

[1] 例句：It involves reducing the amount of memory needed to store a digital image.

分析：本句是简单句。句中 reducing the amount of memory needed to store a digital image 是动名词短语作 involves 的宾语。involve 意为"包含、含有"，后接名词或动名词。例如，a task which involves much difficulty 有困难的任务。

译文：它包括减少存储数字图像所需的存储容量。

[2] 例句：The example below demonstrates how one could go about extracting measurements froman image.

分析：本句是复合句。how one could go about extracting measurements from an image 为宾语从句，作 demonstrates 的宾语。句中 go about 意为从事、干，后跟名词或动名词，例如：Go about your business.忙你自己的事去吧。

译文：下面的例子说明了怎样进行图像测量。

Exercises

Ⅰ. Put "true" or "false" in the brackets for the following statements according to the passage.

1. (　) The digitization procedure can be done by a screen.

2. (　) Image compression involves extracting the amount of memory needed to store a digital image.

3. (　) According to the text, image processing operationgs can be divided into four categories.

4. (　) Once the image is in good condition, the Measurement Extraction operations can be used to obtain useful information from the image.

5. (　) Each pixel in the color is stored as a number between 0 to 255, where 0 represents a black pixel, 255 represents a white pixel and values in-between represent shades of grey.

6. (　) The "clean" image at the right of Figure 7.1 was obtained by applying a median filter to the image.

7. (　) An image with poor contrast, can be improved by reducing the image histogram.

8. (　) Figure 7.3 shows an application of the 2-dimensional Fast Fourier Transform.

9. (　) The aim of Image Measurement Extraction is to extract information about the distribution of the sizes (visible areas) of the objects.

10. (　) Image defects could be caused by the digitization process or by faults in the imaging set-up (for example, bad lighting).

11. (　) Image defects can not be corrected using Image Enhancement techniques.

Ⅱ. Fill in the blanks according to the passage.

1. An image is digitized to convert it to a form that can be stored in a computer's memory or on some form of _____ such as a hard disk or CD-ROM.

2. Once an image has been digitized, it can be operated upon by various _____ operations.

3. Image processing operations can be roughly divided into _____ major categories, _____, Image Enhancement and Restoration, and Measurement Extraction.

4. The digitization procedure can be done by a _____.

5. Image compression involves _____ the amount of memory needed to store a digital image.

6. Image defects can be _____ using Image Enhancement techniques.

7. Once the image is in good condition, the Measurement Extraction operations can be used to _____ useful information from the image.

8. An image with poor contrast can be improved by _____ the image histogram.

9. The "clean" image at the right of Figure 7.1 was obtained by applying a _____ to the image.

10. The image at the top left of Figure 7.3 has a corrugated effect. This can be removed by doing a _____ Fast-Fourier Transform on the image.

III. Translate the following words and phrases into Chinese.

1. storage
2. media
3. category
4. digital image
5. pixel
6. represent
7. corrupt
8. noise
9. background
10. continuous

7.2.1 Reading Material

Can Graphics Files Be Infected with a Virus?

For most types of graphics file formats currently available the answer is "no". A virus (or worm, Trojan horse, and so forth) is fundamentally a collection of code (that is, a program) that contains instructions which are executed by a CPU. Most graphics files, however, contain only static data and no executable code. The code that reads, writes, and displays graphics data is found in translation and display programs and not in the graphics files themselves. If reading or writing a graphics file caused a system malfunction, it is most likely the fault of the program reading the file and not of the graphics file data itself.

With the introduction of multimedia we have seen new formats appear, and modifications to older formats made, which allow executable instructions to be stored within a file format. These instructions are used to direct multimedia applications to play sounds or music, prompt the user for information, or display other graphics and video information.

And such multimedia display programs may perform these functions by interfacing with their environment via an API, or by direct interaction with the operating system. One might also imagine a truly object-oriented graphics file as containing the code required to read, write, and display itself.

Once again, any catastrophes that result from using these multimedia application is most like the result of unfound bugs in the software and not some sinister instructions in the graphics file data. Such "logic bombs" are typically exorcised through the use of testing using a wide variety of different image files for test cases.

If you have a virus scanning program that indicates a specific graphics file is infected by virus, then it is very possible that the file coincidentally contains a byte pattern that the scanning programming recognizes as a key byte signature identifying a virus.

Contact the author (or even read the documentation!) of the virus scanning program to discuss the probability of the mis-identification of a clean file as being infected by a virus. Save the graphics file, as the author will most likely wish to examine it as well.

If you suspect a graphics file to be at the heart of a virus problem you are experiencing, then also consider the possibility that the graphics file's transport mechanism (floppy disk, tape or shell archive file, compressed archive file, and so forth) might be the original source of the virus and not the graphics file itself.

Words

archive	v.	存档
	n.	档案文件
catastrophe	n.	大灾难，大祸
coincidental	adj.	巧合的
exorcise	v.	驱邪，除怪
fault	n.	过错，缺点，故障，毛病
	v.	挑剔，弄错
identification	n.	辨认，证明
indicate	v.	指出，显示，象征，简要地说明
infected	adj.	被感染的
interaction	n.	交互作用，交感
interfacing	n.	界面连接，接口连接
imagine	v.	想像，设想
malfunction	n.	故障
modification	n.	更改，修改，修正
sinister	adj.	险恶的

7.2.2　正文参考译文

数字图像处理介绍

图像可被数字化处理，转换成可存储在计算机存储器或其他存储介质如硬盘或CD-ROM上的格式。这种数字化处理过程可以通过扫描仪或者通过连接到计算机主板上的视频摄像机来实现。当对图像数字化处理后，就可以进行各种图像处理操作。

图像处理操作可以大概分成三大类：图像压缩、图像增强和恢复、测量提取。图像压缩是大多数人都熟悉的。它包括减少存储数字图像所需的存储容量。

利用图像增强技术可以修改数字化处理过程中造成的图像缺陷，或成像时的错误(如灯光不好)造成的图像缺陷。当图像处于好的状况时，就可以利用测量提取操作从图像中得到有用信息。

下面给出了一些图像增强和测量提取的例子。这些例子都是对256级灰度图进行的操作。也就是说，图像中的每个像素以0~255之间的数字存储，0表示黑色像素，255表示白色像素，之间的数值表示渐变的灰色。这些操作也可以扩展到彩色图像。

下面的例子只是给出图像操作技术中的少数几种，没有给出操作的内部工作细节。

图像增强和恢复

图7.1中左边的图在数字化处理过程中被噪声破坏。右边干净的图是通过应用中值滤波器得到的。

图7.2左边的图像对比度差通过调节图像的直方图可以得到右边所示的图像。

图7.3左上角的图由于成像过程中的错误，有波纹影响。这可以通过对图像做二维快

速傅里叶变换来去除(图 7.3 右上角)，去掉亮点(图 7.3 左下角)，最后再通过快速傅立叶反变换得到无波纹的原图像(图 7.3 右下角)。

一个在差的灯光条件下得到的图像，背景亮度显示出连续的变化(图 7.4 的左上角)，可以用下述方法修改。首先通过 25×25 的灰度放大操作去除前景对象(图 7.4 右上角)。然后从背景图像中减去原图像(图 7.4 左下角)。最后翻转颜色并且通过调节直方图改变对比度(图 7.4 右下角)。

图像测量提取

下面的例子说明了怎样进行图像测量。图 7.5 左上角的图像显示出一些物体。目标是要提取物体尺寸(可视部分)分布的信息。第一步包括分割图像，将物体与背景分离。这通常包括为图像限定阈值，在特定阈值之上的像素的值作为白色，其他的作为黑色(图 7.5 的右上角)。因为物体互相接触，设置的阈值包括所有物体的表面，不能显示出分离的物体。这个问题通过对图像实行分水岭分离来实现(图 7.5 左下角)。图 7.5 右下角显示出对左边的两个图像进行逻辑与处理后的结果。这显示出了分水岭分离法对原图中相接触的物体进行分离产生的效果。最后可以从图像中得到测量值。

7.2.3 阅读材料参考译文

图形文件可以感染病毒吗？

对于现有的大多数图形文件格式类型，答案是"否"。病毒(或蠕虫、木马等)基本上是一些代码(即程序)，其中包含由 CPU 执行的指令。但是，大多数图形文件只包含静态数据，不包括可执行的代码。读、写并显示图形数据的代码存在于翻译和显示程序中，而并不存在于图形文件本身。如果读或写图形文件导致系统故障，最可能的原因是读取该文件的程序，而不是图形文件的数据本身。

由于采用多媒体，我们已看到：新的格式出现，旧的格式被修改，即允许可执行指令储存在文件格式内。这些指令用来指示多媒体应用程序播放声音或音乐、提示用户信息，或显示其他图像和视频信息。

而且，这样的多媒体播放程序可以通过用 API 与软件环境连接或直接与操作系统相互作用履行上述职能。我们还可以认为：一个真正的面向对象的图形文件应该包含读、写和显示自己所需要的代码。

使用这些多媒体所造成的灾难很像软件中未被发现的错误引发的后果，而不像图形文件数据中的一些险恶指令。这种"逻辑炸弹"通常可以通过使用各种不同的图像文件进行案例测试去除。

如果你的病毒扫描程序显示特定的图形文件受到病毒感染，那么很可能该文件正好包含一个被扫描程序标识为病毒关键字节特征的字节。

联系病毒扫描程序的作者(或阅读文档！)，看看是不是将干净的文件错误识别为感染病毒的文件了。记得保存好这一图形文件，因为作者也很可能希望检查该图形文件。

如果您怀疑图形文件是您所遇到的病毒问题的核心，还要考虑这种可能性：即病毒的原始来源可能是图形文件的传输机制(软盘、磁带或 shell 档案文件、压缩档案文件等)，而不是图形文件本身。

7.3 Image Compression

Why is image compression so important? Image files, in an uncompressed form, are very large. And the Internet, especially for people using a 56k dialup modem, can be pretty slow.[1] This combination could seriously limit one of the Web's most appreciated aspects—its ability to present images easily.

JPEG (Joint Photographic Experts Group)compression is currently the best way to compress PHOTOGRAPHIC IMAGES for the web. Other forms of image compression, including GIF and PNG, are best used for other purposes on the web.

GIF (Graphics Interchange Format) is best used for graphics that have a limited color pallet and large areas of flat tone, like cartoons or banners. Although it has several remarkable features, such as transparency and the ability to present animated images, it is not well suited for the presentation of continuous tone images, such as photographs, due to its limit of 256 colors.

PNG (Portable Network Graphics) is a relatively new format with a lot of potential but, until all browsers can see images compressed in PNG form, it is not a good idea to use it.

JPEG, or JPG, is an evolving format that is universal in its use as a means of compressing continuous tone photographs for speedy transmission over the Internet. Photographs compressed using the JPEG format look good because JPEG supports millions of colors, so you can see the gradation of tones.

A Bitmap is a simple series of pixels all stacked up. But the same image saved in GIF or JPEG format uses less bytes to make up the file. How? Compression.

"Compression" is a computer term that represents a variety of mathematical formats used to compress an image's byte size. Let's say you have an image where the upper right-hand corner has four pixels all the same color. Why not find a way to make those four pixels into one?[2] That would cut down the number of bytes by three-fourths, at least in the one corner. That's a compression factor.

Bitmaps can be compressed to a point. The process is called "run-length encoding." Runs of pixels that are all the same color are all combined into one pixel. [3] The longer the run of pixels, the more compression. Bitmaps with little detail or color variance will really compress. Those with a great deal of detail don't offer much in the way of compression. Bitmaps that use the run-length encoding can carry either the common ".bmp" extension or ".rle". Another difference between the two files is that the common Bitmap can accept 16 million different colors per pixel. Saving the same image in run-length encoding knocks the bits-per-pixel down to 8. That locks the level of color in at no more than 256.

So, why not create a single pixel when all of the colors are close? You could even lower the number of colors available so that you would have a better chance of the pixels being close in color. Good idea. The people at CompuServe felt the same way.

GIF, which stands for "Graphic Interchange Format," was first standardized in 1987 by

CompuServe, although the patent for the algorithm (mathematical formula) used to create GIF compression actually belongs to Unisys. The first format of GIF used on the Web was called GIF87a, representing its year and version. It saved images at 8 bits-per-pixel, capping the color level at 256. That 8-bit level allowed the image to work across multiple server styles, including CompuServe, TCP/IP.

CompuServe updated the GIF format in 1989 to include animation, transparency, and interlacing. They called the new format, you guessed it: GIF89a.

There's no discernable difference between a basic (known as non-interlaced) GIF in 87 and 89 formats.

Words

banner	n.	旗帜，标语
discernable	adj.	可辨别的，可认识的
evolving	adj.	进化的，展开的
feature	n.	特色
gradation	n.	分等级，层次
means	n.	手段，方法
pallet	n.	盘
remarkable	adj.	显著的
suit	v.	合适
stack	n.	堆，堆栈
	v.	堆叠
tone	n.	色调
transparency	n.	透明，透明度，幻灯片，有图案的玻璃
universal	adj.	普遍的，通用的
variance	n.	不一致，变化

Phrases

animated image	动态图像
run-length encoding	游程长度编码

Abbreviations

GIF (Graphics Interchange Format)	可交换的图像文件格式
JPEG (Joint Photographic Experts Group)	联合图像专家组
PNG (Portable Network Graphics)	便携式网络图像
TCP/IP(Transfer Control Protocol)	传输控制协议/网际协议

Notes

[1] 例句：This combination could seriously limit one of the Web's most appreciated spects

Chapter 7　Image Processing

　　——its ability to present images easily.

　　分析：本句是简单句。句中 its ability to present images easily 为 one of the web's most appreciated aspects 的同位语。

　　译文：这一结合很大程度上限制了网络最令人欣赏的方面——易于显示图像的能力。

[2]　例句：That would cut down the number of bytes by three-fourths, at least in the one corner.

　　分析：本句是简单句。句中 by 意为"到……程度"，例如：The carpet is too short by three feet.地毯短了 3 英尺。

　　译文：那样将会至少在这一个角将字节的数目减少 3/4。

[3]　例句：The longer the run of pixels, the more compression.

　　分析：本句是 "the +比较级，the +比较级" 以省略形式出现的主从复合句，前面部分是从句，后面部分是主句，意为 "越……，越……"。

　　译文：游程越长，压缩幅度越大。

Exercises

Ⅰ. Put "true" or "false" in the brackets for the following statements according to the passage.

1. (　) Image compression is important because uncompressed image files that are very large could limit one of the Web's most appreciated aspects.

2. (　) GIF (Graphics Interchange Format) is not used for graphics that have a limited color pallet and large areas of flat tone, like cartoons or banners.

3. (　) Photographs compressed using the jpeg format look good because jpeg supports millions of colors.

4. (　) A Bitmap is a simple series of pixels all stacked up.

5. (　) GIF stands for "Graphic Internet Format".

6. (　) GIF87a saved images at 8 pits-per-pixel, capping the color level at 256.

7. (　) JPEG, or JPG stands for (Joint Photographic Experts Group).

8. (　) "Compression" is a computer term that represents a variety of mathematical formats used to decompress an image's byte size.

9. (　) A Bitmap is a simple series of pixels all stacked up. But the same image saved in GIF or JPEG format uses more bytes to make up the file.

10. (　) The aim of Compression is to reduce file's size.

Ⅱ. Fill in the blanks according to the passage.

1. Image compression is＿＿＿＿＿＿＿＿＿.

2. Image files, in an ＿＿＿＿＿＿＿＿＿ form, are very large.

3. JPEG compression is currently the best way to＿＿＿＿＿＿＿＿＿ PHOTOGRAPHIC IMAGES for the web.

4. GIF (Graphics Interchange Format) is best used for graphics that have a ＿＿＿＿＿ color pallet and ＿＿＿＿＿ areas of flat tone, like cartoons or banners.

5. GIF is not well suited for the ＿＿＿＿＿＿＿＿＿ of continuous tone images, such as

photographs, due to its limit of 256 colors.

6. PNG (Portable Network Graphics) is a relatively new format with a lot of _____.

7. JPEG supports millions of _____.

8. "Compression" is a computer term that represents a variety of mathematical formats used to _____ an image's byte _____.

9. GIF stands for _____.

10. There's no discernable _____ between a basic (known as non-interlaced) GIF in 87 and 89 formats.

III. Translate the following words and phrases into Chinese.

1. compression
2. uncompress
3. combination
4. seriously
5. ability
6. format
7. byte
8. mathematical
9. extension
10. bits-per-pixel

7.3.1 Reading Material

Introduction to the Image Compression Manager

The Image Compression Manager provides your application with an interface for compressing and decompressing images and sequences of images that is independent of devices and algorithms.

Uncompressed image data requires a large amount of storage space. Storing a single 640-by-480 pixel image in 32-bit color can require as much as 1.2 MB. Sequences of images, like those that might be contained in a QuickTime movie, demand substantially more storage than single images. This is true even for sequences that consist of fairly small images, because the movie consists of such a large number of those images. Consequently, minimizing the storage requirements for image data is an important consideration for any application that works with images or sequences of images.

The Image Compression Manager allows your application to

- use a common interface for all image-compression and image-decompression operations;
- take advantage of any compression software or hardware that may be present in a given Macintosh configuration;
- store compressed image data in pictures;
- temporarily compress sequences of images, further reducing the storage requirements of movies;
- display compressed PICT files without the need to modify your application;
- use an interface that is appropriate for your application—a high-level interface if you do not need to manipulate many compression parameters or a low-level interface that provides you greater control over the compression operation.

The Image Compression Manager compresses images by invoking image compressor components and decompresses images using image decompressor components. Compressor and decompressor components are code resources that present a standard interface to the Image Compression Manager and provide image-compression and image-decompression services, respectively. The Image Compression Manager receives application requests and coordinates the actions of the appropriate components. The components perform the actual compression and decompression. Compressor and decompressor components are standard components and are managed by the Component Manager.

Because the Image Compression Manager is independent of specific compression algorithms and drivers, it provides a number of advantages to developers of image- compression algorithms. Specifically, compressor and decompressor components can

- present a common application interface for software-based compressors and hardware-based compressors;
- provide several different compressors and compression options, allowing the Image Compression Manager or the application to choose the appropriate tool for a particular situation.

Words

appropriate	*adj.*	适当的
coordinate	*n.*	相配之物
consequently	*adv.*	从而，因此
decompression	*n.*	解压
decompressor	*n.*	解压程序
demand	*n.*	要求，需要
	v.	要求，查询
invoke	*v.*	调用
minimizing	*v.*	极小化，求最小参数值
request	*v./ n*	请求，要求
substantially	*adv.*	充分地
Macintosh (Mac)	*n.*	苹果公司生产的一种型号的计算机

7.3.2 正文参考译文

图像压缩

为什么图像压缩如此重要？因为非压缩格式的图像文件非常大。而 Internet，尤其对使用 56k 调制解调器拨号上网的用户来说，是相当慢的。巨大的图像和慢速上网的结合很大程度上限制了网络最令人欣赏的方面，即易于显示图像的能力。

JPEG(联合图像专家组)压缩是当前 Web 压缩图像的最好方法，其他图像压缩格式，包括 GIF 和 PNG，更多地用于 Web 外的其他目的。

GIF(可交换的图像文件格式)最适合于色彩较少并且有大面积的均匀色调的图像，如卡通或旗帜。尽管它有几个明显的特点，譬如透明以及显示动态图像的能力，但是由于 256 色的限制，它不适于显示照片之类色调连续的图像。

PNG(便携式网络图像)是一种相对较新的有很大潜力的格式。但是，在所有浏览器可以浏览 PNG 格式的图像之前，最好不要使用该格式。

JPEG 或 JPG 格式是一个发展中的通用格式，用于压缩连续色调图像，以提高网络传输速度。采用 JPEG 格式压缩的照片看起来很好，因为 JPEG 支持数百万种色彩，所以我们可以看到色调的分级。

位图文件是一系列的像素堆积起来的。但是同一幅图片用 GIF 或 JPEG 格式保存只用较少的字节即可形成文件。怎样实现？就是用压缩的方法。

压缩是一个计算机术语，表示用于压缩图像字节数的数学格式。假设有一幅图像的右上角有四个同色的像素，为什么不找一个方法将四个像素放在一起？那样将会至少在该区域将字节的数目减少 3/4。这就是一个压缩因子。

位图可以被压缩成一个点，这个过程叫做"行程编码算法"。同一行上所有颜色相同的相邻都可以用同一个像素表示，这些颜色相同的像素之间的距离越大，压缩的幅度就越大。细节和色彩变化较少的位图的确可以压缩，那些有很多细节的位图用这种方法没有太大的压缩空间。使用游程长度编码的位图文件扩展名可以是".bmp"或".rle"。这两种文件的另一区别是普通位图文件可以接收每像素 16 000 000 种不同的颜色，采用游程长度编码保存同一个图像将每像素字节数减少到 8，这限制颜色分级不超过 256。

那么，当所有颜色相近的时候为什么不创建一个单一的像素呢？我们甚至可以降低可用颜色的数量，以使得像素能得到相近的颜色。这是个好办法，CompuServe 公司的人也是这样想的。

GIF(即 Graphic Interchange Format，可交换的图像文件格式)于 1987 年由 CompuServe 公司给出标准，而用来产生 GIF 压缩的算法(数学公式)的专利属于 Unisys。用于 Web 的第一个 GIF 格式叫做 GIF87a，以表明其年份和版本。这种格式以每像素 8 个二进制位存储图像，颜色分级的上限为 256。这种 8 位分级允许图像工作于多种服务器风格，包括 CompuServe、TCP/IP。

1989 年，CompuServe 公司更新了 GIF 格式，使其可以包含动画、透明和交织。你可以猜测到：他们称其新格式为 GIF89a。

87 和 89 格式的基本(即无交织的)GIF 无明显不同。

7.3.3　阅读材料参考译文

图像压缩管理器的介绍

图像压缩管理器为应用程序提供接口，用于压缩和解压不受设备和算法影响的图像以及图像序列。

未压缩的图像数据需要大量的存储空间。在 32 位彩色中存储一个单一的 640×480 像素的图像可能需要多达 1.2MB 的空间。图像序列——与 QuickTime 影片包含的图像序列一样，需要比单一图像存储更多的储存空间。对于相当小的图像序列也是如此，因为影片包括数量非常多的图像。所以，对于任何使用图像或图像序列工作的应用程序来说，尽量地减少图像数据的存储需求是需要着重考虑的问题。

图像压缩管理器允许应用程序
- 所有的图像压缩和图像解压操作使用一个通用接口；
- 利用可能存在于 Macintosh 电脑配置中的压缩软件或硬件；
- 在图片中存储压缩图像数据；
- 临时压缩图像序列，进一步减少电影的存储需求；
- 显示压缩 PICT 文件，而无需修改应用；
- 使用适合应用程序的接口——如果不需要操纵许多压缩参数，则使用高级接口；或者使用低级接口，以对压缩操作提供更多的控制。

图像压缩管理器通过调用图像压缩器组件压缩图像，并通过使用图像解压缩器组件解压图像。压缩器及解压器组件是代码资源，该资源向图像压缩管理器提供一个标准接口，并分别提供图像压缩和图像解压服务。图像压缩管理器接受应用请求，并协调适当部件的动作。这些部件执行实际压缩和解压。压缩器及解压器部件属于标准组件，由组件管理器管理。

由于图像压缩管理器不受特定的压缩算法和驱动程序的影响，所以它向图像压缩算法的设计者提供了很多便利。具体来说，压缩器及解压器组件能够做以下事情：
- 为基于软件的压缩器和基于硬件的压缩器提供共同的应用界面；
- 提供几种不同的压缩器和压缩选择，允许图像压缩管理器或应用程序为特定的情形选择适当的工具。

7.4 Application of Digital Image Processing

The field of digital image processing has experienced continuous and significant expansion in recent years. The usefulness of this technology is apparent in many different disciplines covering medicine through remote sensing. The advances and wide availability of image processing hardware has further enhanced the usefulness of image processing.

Remote sensing is the process of collecting data about objects or landscape features without coming into direct physical contact with them.

Digital Image Processing is not only a step in the remote sensing process, but is itself a process that consists of several steps. It is important to remember that the ultimate goal of this process is to extract information from an image that is not readily apparent or is not available in its original form. The steps taken in processing an image will vary from image to image for multiple reasons, including the format and initial condition of the image, the information of interest (i.e., geologic formations vs. land cover), the composition of scene elements. There are three general steps in processing a digital image: preprocessing, display and enhancement, and information extraction.

Preprocessing—Before digital images can be analyzed, they usually require some degree of preprocessing. This may involve radiometric corrections, which attempt to remove the effects of sensor errors and/or environmental factors.[1] A common method of determining what errors have been introduced into an image is by modeling the scene at the time of data acquisition using ancillary data collected.

Geometric corrections are also very common prior to any image analysis. If any types of area, direction or distance measurements are to be made using an image, it must be rectified if they are to be accurate.[2] Geometric rectification is a process by which points in an image are registered to corresponding points on a map or another image that has already been rectified. The goal of geometric rectification is to put image elements in their proper planimetric (x and y) positions.

Information Enhancement—There are numerous procedures that can be performed to enhance an image. However, they can be classified into two major categories: point operations and local operations. Point operations change the value of each individual pixel independent of all other pixel, while local operations change the value of individual pixels in the context of the values of neighboring pixels. Common enhancements include image reduction, image magnification, transect extraction, contrast adjustments (linear and non-linear), band ratioing, spatial filtering, fourier transformations, principle components analysis, and texture transformations.

Information Extraction—Unlike analog image processing, digital image processing presently relies almost wholly on the primary elements of tone and color of image pixels.

There has been some success with expert systems and neural networks which attempt to enable the computer to mimic the ways in which humans interpret images. Expert systems accomplish this through the compilation of a large database of human knowledge gained from analog image interpretation which the computer draws upon in its interpretations.[3] Neural networks attempt to "teach" the computer what decisions to make based upon a training data set. Once it has "learned" how to classify the training data successfully, it is used to interpret and classify new data sets.

Once the remotely sensed data has been processed, it must be placed into a format that can effectively transmit the information it was intended to. This can be done in a variety of ways including a printout of the enhanced image itself, and image map, a thematic map, a spatial database, summary statistics and/or graphs. Because there are a variety of ways in which the output can be displayed, a knowledge not only of remote sensing, but of such fields as GIS, cartography, and spatial statistics is a necessity. With an understanding of these areas and how they interact one with another, it is possible to produce output that gives the user the information needed without confusion. However, without such knowledge it is more probable that output will be poor and difficult to use properly, thus wasting the time and effort expended in processing the remotely sensed data.

Words

ancillary	*adj.*	补助的，副的
cartography	*n.*	绘图法
compilation	*n.*	编辑，编制，复杂

confusion	n.	混乱，混淆
discipline	n.	学科
extract	n.	榨出物
	v.	析取
features	n	容貌，特征
interpretation	n.	解释，通译
landscape	n.	地形
magnification	n.	扩大，放大倍率
mimic	adj.	模仿的
	v.	摹拟
necessity	n.	必要性，必需品
neural	adj.	神经系统的
numerous	adj.	众多的，许多的，无数的
planimetric	adj.	平面的，地球表面经纬度的
preprocess	v.	预加工，预处理
processing	n.	处理
properly	adv.	适当地，完全地
radiometric	adj.	辐射测量的
rectification	n.	纠正，整顿，校正
reduction	n.	减少，缩减量
texture	n.	(织品的)质地，(木材等的)纹理
thematic	adj.	主题的
transect	v.	横断
	n.	横断面

phrases

expert systems	专家系统
neural networks	神经网络
remote sensing	遥感，遥测
geologic formations	地质层
land cover	土地覆盖

Abbreviations

GIS(Geographic Information System)	地理信息系统

Notes

[1] 例句：A common method of determining what errors have been introduced into an image is by modeling the scene at the time of data acquisition using ancillary data collected.

分析：本句为复合句。句中 of determining what errors have been introduced into an image 为 of+动名词短语，构成介词短语作定语修饰 a common method。by modeling the scene at the time of data acquisition using ancillary data collected 是介词短语作表语。

译文：判定什么错误已经传入图像的常用方法是利用收集到的辅助数据对获得数据时的景象进行建模。

[2] 例句：Geometric rectification is a process by which points in an image are registered to corresponding points on a map or another image that has already been rectified.

分析：本句是复合句，句中 by which 引出的定语从句修饰 process。

译文：几何校正是图像中的点与地图或另一个已校准的图像中相应的点配准的过程。

[3] 例句：Neural networks attempt to "teach" the computer what decisions to make based upon a training data set.

分析：本句是简单句。句中 what decisions to make 为疑问词+不定式结构，在句中作 teach 的直接宾语。这种结构也可称为不定式短语，在句中可作主语、表语、宾语等。举例如下：

主语　　How to operate a computer is known to the students.
表语　　The important thing is where to get the necessary information.
宾语　　Please tell me what to do next time.

译文：神经网络试图"教"给计算机基于一组训练数据做出判断。

Exercises

Ⅰ. Put "true" or "false" in the brackets for the following statements according to the passage.

1. (　) The usefulness of digital image processing is apparent in medicine field.
2. (　) Remote sensing is an example of digital image processing.
3. (　) Remote sensing is the process of collecting data about objects or landscape features by coming into direct physical contact with them.
4. (　) Digital Image Processing is a process that consists of several steps.
5. (　) The steps taken in processing an image are the same with each image.
6. (　) There are three general steps in processing a digital image; preprocessing, display and enhancement, and information extraction.
7. (　) Geometric corrections are useless.
8. (　) Information Enhancement can be classified into two major categories: point operations and local operations.
9. (　) Once the remotely sensed data has been processed, it must be placed into a format that can effectively transmit the information it was intended to.
10. (　) Digital image processing technology can not be used in medicine.

Ⅱ. Fill in the blanks according to the passage.

1. The advances and wide availability of image processing hardware has further enhanced

Chapter 7 Image Processing

the usefulness of _____.

2. Remote sensing is the process of _____ data about objects or landscape features without coming into direct physical contact with them.

3. The ultimate goal of image processing is to _____ information from an image that is not readily apparent or is not _____ in its original form.

4. There are three general steps in processing a digital image: _____, display and enhancement, and information extraction.

5. The goal of _____ rectification is to put image elements in their proper planimetric (x and y) positions.

6. Point operations change the value of each _____ pixel independent of all other pixels.

7. _____ change the value of individual pixels in the context of the values of neighboring pixels.

8. Digital image processing presently relies almost wholly on the _____ elements of tone and color of image pixels.

9. There has been some success with _____ and neural networks which attempt to enable the computer to _____ the ways in which humans interpret images.

10. Neural networks attempt to "teach" the computer what _____ to make based upon a training data set.

III. Translate the following words and phrases into Chinese.

1. application
2. Digital Image Processing
3. extract
4. multiple
5. enhancement
6. extraction
7. correction
8. expert system
9. classify
10. interact

7.4.1 Reading Material

Graphics and Computer-Aided Design (CAD)

Graphics is the language of communications for engineers, designers, and architects. Technical drawings are the means for describing something that must be processed, manufactured, or built. Engineers, designers, and architects use technical drawings as a means of communicating their ideas. Until the 1950s and the advent of the computer, technical drawings were done at the drafting table with paper, pencil, and T–squares. Now most technical drawings are done at the computer.

Computer-aided design (CAD) refers to a system that uses computers with advanced graphics hardware and software to create precision drawings or technical illustrations. If the system is being used to design parts to be manufactured, the designer can draw and manipulate a 3-D image of the part without having to build a physical model. If the system is being used to produce architectural drawings, the structure can be drawn and viewed from different perspectives. Although CAD systems can be thought of as simulating the paper, pencil, and T-square, they have far more complex capabilities.

CAD systems can be broadly classified as two-dimensional (2-D) CAD and three-dimensional (3-D) CAD. Two-dimensional CAD systems are basically glorified electronic drawing boards, replacing paper, pencil and the T-square. Of course, the drawings are easy to edit and reproduce, guaranteeing top-quality copies.

Three-dimensional CAD is also called geometric modeling. There are three methods of modeling in three dimensions: wireframe modeling, surface modeling, and solid modeling.

Wireframe, the simplest 3-D modeling, represents objects by line elements that provide exact information about edges, corners, and surface discontinuities. With these models, there is no way to distinguish between the inside and the outside of the object. Surface modeling, on the other hand, defines precisely the outside of the object being modeled. Surface models connect various types of surface elements by line segments. Solid models make use of topology, the interior volume and mass of an object is defined. Surface models appear similar to solid models, but the interior of the surface model is empty.

Every CAD system has a set of elements or primitives out of which the designs are created. In a 2-D system, the primitives are points, lines and surfaces. Surfaces can be polygons with any number of vertices, they can also be figures such as circles and ellipses. Splines, free curves defined mathematically, are also often included. In 3-D systems, the primitive shapes are cubes, wedges, cylinders, or spheres. In surface modeling, a cube would be composed of six faces, but in solid modeling, a cube would be a single primitive.

CAD systems that are specific to particular types of design may have a set of specialized primitives. An architectural CAD might have architectural components such as slabs, walls, doors, and windows. A CAD system used in designing cars might have components such as bumpers, windshields, and tires.

Even if CAD systems are used only as electronic drawing boards, they provide great advantages. They are much faster and more accurate than hand drawing. Revisions are easier to make, because the unchanged parts do not need to be copied again. If the CAD system is being used as a real design tool, the designer can try out ideas and immediately see the results. "What if" questions can be applied to the model to test the integrity of new designs. If the output of the design is the specification for an item to be manufactured, the specifications can be sent directly to the manufacturing machine.

Of course, the use of CAD has introduced its own set of new problems. Computer-aided design is a concept implemented in many diverse software programs. The designers must learn to use the new software tools, sometimes taking as long as six months to become proficient.

Words

architect	n.	建筑师
bumper	n.	(汽车前后的)保险杠

cylinder	n.	圆筒，圆柱体，汽缸，柱面
diverse	adj.	不同的，变化多的
drawing	n.	图画，制图，素描术
ellipse	n.	[数]椭圆，椭圆形
illustration	n.	说明，例证，例子，图表，插图
interior	adj.	内部的，内的
	n.	内部
means	n.	手段，方法
polygon	n.	[数]多角形，多边形
proficient	n.	精通
	adj.	(常与 in、at 连用)熟练的；精通的
slab	n.	厚平板，混凝土路面，板层
sphere	n.	球，球体，范围，领域，方面，半球
spline	n.	方栓，齿条
t–square	n.	丁字尺
tire	v.	使……疲倦，厌烦
	n.	轮胎
vertices	n.	至高点，头顶，
vertex	n.	顶角，(角的)顶点
wedge	n.	楔子，三角木，楔形物
windshield	n.	(汽车)挡风玻璃

Abbreviations

CAD (computer-aided design)　　　　　　计算机辅助设计

7.4.2　正文参考译文

数字图像处理的应用

近年来，数字图像处理领域不断扩大，从医学领域到遥感技术，数字图像处理技术在很多学科都得到了显著的应用。图像处理硬件的发展和广泛的适应性也进一步促进了图像处理的应用。

遥感技术可以远距离地采集对象或地形特征的数据，而不需要物质的直接接触。

数字图像处理不仅是遥感处理过程中的一个环节，它本身也包括几个步骤，该过程的最终目的是从原始形式不明显或不可用的图像中提取信息。处理图像所采取的步骤根据不同的图像会有所不同，这是因为存在多个原因，包括：图像的格式和初始条件不同、对信息的兴趣不同(例如地质层对比土地覆盖)以及环境成分不同。处理数字图像一般有三个步骤：预处理、显示和增强、信息提取。

预处理——在分析数字图像之前，经常需要对图像进行某种程度的预处理。包括辐射校正，采用这种方法的目的是去除传感器错误和环境因素的影响。判定什么错误已经传入图像的常用方法是利用收集到的辅助数据对获得数据时的景象进行建模。

几何校正也是常用的优于其他图像分析的方法。如果需要利用图像测量区域、方向或距离，需要对它们进行精确调整。几何校正是图像中的点与地图或另一个已校准的图像中相应的点配准的过程。几何校正的目的是将图像元素放在适当的平面位置。

信息增强——有几种方法可以增强图像。这些方法可以分成两大类：点运算和局部运算。点运算独立地改变每一个像素的值，而局部运算根据相邻像素的值改变每一个像素值。常用的增强方法包括图像缩小、图像放大、横断面提取、对比度调节(线形或非线形)、带比调节、空间滤波器、傅里叶变换、主成分分析和纹理转换。

信息提取——与模拟图像处理不同，当前的数字图像处理几乎仅依赖于像素的色调和颜色这些基本要素。

专家系统和神经网络试图让计算机模拟人类理解图像的方式，已经获得了一些成功。专家系统通过编辑模拟图像解译得到的大型人类知识库，由计算机利用图像解译实现模拟。神经网络试图"教"给计算机基于一组训练数据做出判断，当计算机成功地学会了怎样分类训练数据，就可以用来理解和分类新的数据组。

处理遥感数据后，需要把这些数据设置成能有效传输信息的格式。这可以通过几种方法实现，包括打印出增强图像本身、图像映射图、专题地图、空间数据库、摘要统计和图表。因为有多种方式可以显示输出，这就会用到遥感知识以及地理信息系统、绘图法和空间统计等领域的知识。理解了这些领域及其交叉内容，就可以清晰地输出用户需要的信息。无论如何，如果没有这些知识，输出会很差并且很难得到适当的应用，这样就浪费了花费在遥感数据处理上的时间和努力。

7.4.3 阅读材料参考译文

制图学与 CAD

图形是工程师、设计师和建筑师沟通的语言。技术图纸用来描述需要处理、制造、或建造的东西。工程师、设计师和建筑师使用技术图纸作为手段互相交流想法。在 20 世纪 50 年代计算机出现之前，技术图纸是在制图桌上利用纸、铅笔和丁字尺来制作的。现在大部分技术图纸都是在计算机上制作的。

计算机辅助设计(CAD)涉及利用先进的计算机图形硬件和软件以创建精确图纸或技术插图的系统。如果应用该系统设计制造部件，设计人员可以利用和操纵该部件的三维图像，而无需建立物理模型。如果系统用于制作建筑图纸，则可以从不同的角度绘制和观察结构。虽然 CAD 系统可以被看作是模拟纸、铅笔和丁字尺，但它们具有更为复杂的功能。

CAD 系统大致可划分为二维(2-D)CAD 和三维(3-D)CAD。二维 CAD 系统基本上是美化的电子绘图板，取代纸张、铅笔和丁字尺。当然，电子图很容易被编辑和复制，保证是最高质量的副本。

三维 CAD 也被称为几何造型。在三个维度中有三种建模方法：线框模型、曲面模型和实体模型。

线框模型，最简单的 3-D 建模，用线要素描述物体，提供边、角和表面的不连续性的准确信息。使用这些模型，没有办法区分物体的内部与外部。而曲面造型却准确地界定被建模的物体的外面。曲面模型用线段连接不同类型的曲面元素。实体模型利用拓扑，确定一个物体内部的体积和质量。曲面模型看起来类似于实体模型，但曲面模型的内部是空的。

每个 CAD 系统有一套要素或原始设计，利用它们产生新的设计。在 2-D 系统，原始设计是点、线、面。面可以是任意多边形，也可以是圆或椭圆图形。还常常包括样条(即数学定义的自由曲线)。在 3-D 系统，原始的形状是立方体、楔形、圆柱体或半球形。在曲面模型中，立方体由 6 个面组成，但在实体模型中，立方体是一个原始形状。

专用于特殊类型设计的 CAD 系统会有一组专门的原始形状。建筑设计 CAD 会有建筑构件，如砖、墙壁、门和窗户。用于设计汽车的 CAD 系统中的部件可能有保险杠、挡风玻璃和轮胎。

即使 CAD 系统仅作为电子绘图板，它们仍具有很大的优势。它们比手工绘图更快、更准确。修改更容易，因为无改动的部分不需要再次复制。如果使用 CAD 系统作为一个真正的设计工具，设计人员可以尝试新想法，并立即看到结果。"如果……会怎样"的问题可应用于模型以测试新设计的完整性。如果设计结果是某个要生产的产品的规格，则该规格可以直接送往生产机器。

当然，使用计算机辅助设计引起了一些新的问题。计算机辅助设计是应用于许多不同的软件程序的概念。设计者必须学会使用新的软件工具，有时需要长达六个月的时间才能精通该工具。

Chapter 8 Embedded Systems

8.1 Come to Study Embedded Systems

An embedded system is a special-purpose computer system designed to perform one or a few dedicated functions, often with real-time computing constraints. It is usually embedded as part of a complete device including hardware and mechanical parts. In contrast, a general-purpose computer, such as a personal computer, can do many different tasks depending on programming. Embedded systems control many of the common devices in use today. Figure 8.1 shows an example of an embedded system.

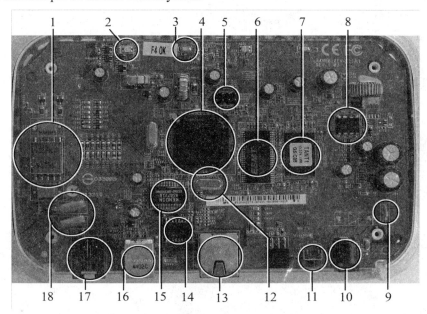

Figure 8.1 Picture of the internals of a Netgear ADSL modem/router. A modern example of an embedded system. Labelled parts include a microprocessor (4), RAM (6), and flash memory (7).

[1]Since the embedded system is dedicated to specific tasks, design engineers can optimize it, reducing the size and cost of the product, or increasing the reliability and performance. Some embedded systems are mass-produced, benefiting from economies of scale.

Physically, embedded systems range from portable devices such as digital watches and MP3 players, to large stationary installations like traffic lights, factory controllers, or the systems controlling nuclear power plants. [2]Complexity varies from low, with a single microcontroller chip, to very high with multiple units, peripherals and networks mounted inside a large chassis or enclosure.

In general, "embedded system" is not an exactly defined term, as many systems have some element of programmability. For example, Handheld computers share some elements with embedded systems—such as the operating systems and microprocessors which power them—but are not truly embedded systems, because they allow different applications to be loaded and peripherals to be connected.

It's important to note that an embedded system in no way implies a real-time system. A real-time system responds to unpredictable external stimuli in a timely and predictable way.

One other interesting distinction of most embedded systems is that the software that is developed for them is built (compiled and linked) on another computer. When we build software on a standard desktop PC, we compile the source and run it on the same machine. In an embedded system, we commonly "cross-compile" our code on one system (the host) and then execute on our embedded system (the target). The host and target commonly differ by architecture, so the cross-compiler generates code that is not executable on the host.

Words

constraint	n.	限制
dedicated	adj.	专用的
embedded	adj.	嵌入式的
optimize	v.	优化
reliability	n.	可靠性
performance	n.	性能
installations	n.	安装，装置
peripheral	n.	外部设备
programmability	n.	可编程序性

Notes

[1] 例句：Since the embedded system is dedicated to specific tasks, design engineers canoptimize it, reducing the size and cost of the product, or increasing the reliability and performance.

分析：本句为复合句。主句中的 reducing the size and cost of the product, or increasing the reliability and performance 为并列的现在分词短语，作结果状语。

译文：由于嵌入式系统专用于具体任务，设计工程师可以对其进行优化，缩减产品的尺寸和成本，或者提高其可靠性和性能。

[2] 例句：Complexity varies from low, with a single microcontroller chip, to very high with multiple units, peripherals and networks mounted inside a large chassis or enclosure.

分析：本句为简单句。句中的 vary from low(A) to very high(B)，意思是在 low(A) 到 very high(B)之间的范围变化。 例如：These fish vary in weight from 3 lb to 5 lb. 这些鱼的重量从 3 磅到 5 磅不等。注意区别：vary from A to A，意思是因 A 不同而不一样。 例如：Opinions vary from person to person. 观点因

人而异。

译文：嵌入式系统的复杂性不一，复杂性低的只有一个微控制器芯片，复杂性非常高的具有安装在大底盘上或封装起来的多个部件、外围设备和网络。

Exercises

Ⅰ. Put "true" or "false" in the brackets for the following statements according to the passage.

1. (　) An embedded system is a special-purpose computer system designed to perform one or a few dedicated functions.
2. (　) An embedded system is usually embedded as part of a complete device including software and mechanical parts.
3. (　) Physically, embedded systems range from portable devices to large stationary installations.
4. (　) Design engineers can optimize "embedded system" to reduce the size and cost of the product, or increase the reliability and performance.
5. (　) In general, "embedded system" is an exactly defined term, as many systems have some element of programmability.
6. (　) Handheld computers share some elements with embedded systems, so they are truly embedded systems.
7. (　) The embedded system is dedicated to general tasks.
8. (　) A general-purpose computer can do many different tasks depending on programming.
9. (　) In an embedded system, we commonly "cross-compile" our code and then execute on the same system.
10. (　) A real-time system responds to unpredictable external stimuli in a timely and predictable way.

Ⅱ. Fill in the blanks according to the passage.

1. An embedded system is a special-purpose computer system designed to perform one or a few _____ functions, often with real-time computing _____.
2. In contrast, a general-purpose computer can do many different tasks depending on _____.
3. Some embedded systems are _____, benefiting from economies of scale.
4. Embedded systems control many of the common _____ in use today.
5. Handheld computers allow different _____ to be loaded and _____ to be connected
6. The _____ is dedicated to specific task.
7. An embedded system is a _____ system designed to perform one or a few dedicated functions, often with real-time computing constraints.
8. In contrast, a _____ can do many different tasks depending on programming.
9. A _____ responds to unpredictable external stimuli in a timely and predictable way.
10. When we build software on a standard desktop PC, we compile the source and run it on the same machine.

III. Translate the following words and expressions into Chinese.

1. embedded system
2. economies of scale
3. peripherals
4. real-time system
5. general-purpose computer
6. real-time computing constraints
7. microprocessors
8. special-purpose
9. cross-compile
10. external stimuli

8.1.1 Reading Material

History of Embedded System

In the earliest years of computers in the 1930~1940s, computers were sometimes dedicated to a single task, but were far too large and expensive for most kinds of tasks performed by embedded computers of today. Over time however, the concept of programmable controllers evolved from traditional electromechanical sequencers, via solid state devices, to the use of computer technology.

One of the first recognizably modern embedded systems was the Apollo Guidance Computer, developed by Charles Stark Draper at the MIT Instrumentation Laboratory. At the project's inception, the Apollo guidance computer was considered the riskiest item in the Apollo project as it employed the then newly developed monolithic integrated circuits to reduce the size and weight. An early mass-produced embedded system was the Autonetics D-17 guidance computer for the Minuteman missile, released in 1961. It was built from transistor logic and had a hard disk for main memory. When the Minuteman II went into production in 1966, the D-17 was replaced with a new computer that was the first high-volume use of integrated circuits. This program alone reduced prices on quad nand gate ICs from $1000/each to $3/each, permitting their use in commercial products.

Since these early applications in the 1960s, embedded systems have come down in price and there has been a dramatic rise in processing power and functionality. The first microprocessor for example, the Intel 4004, was designed for calculators and other small systems but still required many external memory and support chips. In 1978 National Engineering Manufacturers Association released a "standard" for programmable microcontrollers, including almost any computer-based controllers, such as single board computers, numerical, and event-based controllers.

As the cost of microprocessors and microcontrollers fell, it became feasible to replace expensive knob-based analog components, such as potentiometers and variable capacitors, with up/down buttons or knobs read out by a microprocessor even in some consumer products. By the mid-1980s, most of the common previously external system components had been integrated into the same chip as the processor and this modern form of the microcontroller allowed an even more widespread use, which by the end of the decade were the norm rather than the exception for almost all electronics devices.

The integration of microcontrollers has further increased the applications for which embedded systems are used into areas where traditionally a computer would not have been considered. A general purpose and comparatively low-cost microcontroller may often be programmed to fulfill the same role as a large number of separate components. Although in this context an embedded system is usually more complex than a traditional solution, most of the complexity is contained within the microcontroller itself. Very few additional components may be needed and most of the design effort is in the software. The intangible nature of software makes it much easier to prototype and test new revisions compared with the design and construction of a new circuit not using an embedded processor.

Words

capacitor	n.	电容器
electromechanical	adj.	机电的
feasible	adj.	可行的，切实可行的
functionality	n.	功能性，泛函性
inception	n.	起初，获得学位
instrumentation	n.	仪表化，使用仪器
intangible	adj.	无形的
	adj.	难以明了的，无形的
microcontroller	n.	微控制器
Minuteman	n.	民兵(导弹名)
monolithic	adj.	完全统一的，整体的
	n.	单片电，单块集成电路
potentiometers	n.	电位器，分压计
prototype	n.	原型

8.1.2　正文参考译文

开始学习嵌入式系统

嵌入式系统是一种特殊目的的计算机系统，被设计用来执行一个或几个专门的功能，往往只用于实时计算。通常作为包括硬件和机械部件的完整设备的一部分嵌入设备中。而通用计算机，如个人电脑，可以根据编程的不同执行许多不同的任务。嵌入式系统控制用于现在的许多常见装置中。图8.1是一个嵌入式系统的例子。

由于嵌入式系统专用于具体任务，设计工程师可以对其进行优化，缩减产品的尺寸和成本，提高其可靠性和性能。得益于经营规模扩大，一些嵌入式系统正在大量生产。

嵌入式系统的应用范围从便携式设备如数字手表和MP3播放器，到大型固定装置如交通灯、工厂控制器、核电厂控制系统，应有尽有。嵌入式系统的复杂性不一，复杂性低的只有一个微控制器芯片，复杂性非常高的具有安装在大底盘上或封装起来的多个部件、外围设备和网络。

通常，术语"嵌入式系统"定义得并不够准确，因为许多系统有一些可编程能力。例如，掌上电脑与嵌入式系统有一些共同点——如操作系统以及为操作系统提供支持的微处理器——但不是真正的嵌入式系统,因为它们允许加载不同的应用程序和连接外部设备。

需要重点注意的是:嵌入式系统绝不意味着实时系统。实时系统以及时、可预测的方式对不可预测的外部刺激做出反应。

大多数嵌入式系统的另一个特点是:为嵌入式系统开发的软件可在另一台计算机上生成(编译和链接)。我们在标准台式电脑上生成软件后,在同一台机器编译并运行源代码。在嵌入式系统中,常把代码在一个系统机(主机)上"交叉编译",然后在嵌入式系统(目标机)上执行。主机和目标机通常是不同架构的,所以交叉编译生成的代码不能在主机上执行。

8.1.3 阅读材料参考译文

嵌入式系统的历史

二十世纪三四十年代,在计算机诞生以后的最初几年里,计算机往往仅执行单一的任务,相对于今天的嵌入式计算机所完成的大多数任务来说,那时的计算机过于庞大和昂贵。但是,随着时间的推移,可编程控制器的概念已从传统的电动机械的序列发生器经过固态装置的控制器发展到了计算机技术的应用。

阿波罗制导计算机是公认的最早的现代嵌入式系统之一,由查尔斯·斯塔克·德雷珀在麻省理工学院德雷珀实验室研制而成。在阿波罗计划开始的时候,阿波罗制导计算机被认为是该计划中风险最大的项目,因为它采用了当时新开发的单片集成电路来减小计算机的尺寸,减轻其重量。早期大规模生产的嵌入式系统是用于"民兵导弹"的 Autonetics D-17 制导计算机,它在 1961 年发售。D-17 由晶体管逻辑电路组成,有一个硬盘做主存。当民兵Ⅱ在 1966 年投产时,D-17 被首次大量使用集成电路的新电脑代替。仅这一计划已将四与非门集成电路的价格从每个 1000 美元降至每个 3 美元,使得其可以在商品中使用。

嵌入式系统在 20 世纪 60 年代开始应用。之后,其价格不断下降,处理能力和性能却大幅增强。例如,第一个微处理器 Intel 4004 是专为计算机和其他小系统设计的,但仍需要许多外部存储器和支持芯片。1978 年,美国国家工程制造商协会发布了可编程微控制器的"标准",该"标准"几乎可针对所有的以计算机为基础的控制器,如单板计算机、数控以及基于事件的控制器。

随着微处理器和微控制器成本的下降,在一些消费产品中,由微处理器可读取的上/下按钮或旋钮取代昂贵的以旋钮为基础的部件(如电位器和可变电容器之类)成为可能。到 20 世纪 80 年代中期,大多数以前常见的外部系统组件已与处理器集成到同一个芯片,这种现代形式的微控制器得到了更为广泛的应用。到 20 世纪 80 年代末,这种整合在所有电子设备中都几乎可以看到。

微控制器的集成进一步扩大了嵌入式系统在一些领域的应用,这些领域传统上是不考虑应用计算机的。人们经常对通用和成本较低的微控制器编程,使之发挥与大量分立元件相同的作用。尽管在这种情况下,嵌入式系统通常比传统的解决方案更复杂,但其复杂性主要在微控制器本身。大部分设计工作都包含在软件里,不需要附加组件。与设计和制造不采用嵌入式处理器的新电路相比,软件的无形性使得设计和测试采用嵌入式处理器的新方案更容易。

8.2　Characteristics of Embedded Systems

The key characteristic of an embedded system is that it is supposed to handle a few simple tasks, although the steps involved in handling or accomplishing that task may be as complex as any computer program. A videogame controller, for example, may be said to have simple tasks - load the game and allow the player to control it through commands entered through the handset. In truth, however, a game controller (especially the newer games built for the X-box or PS3) goes through a series of steps and actions that require as much processing power as a standalone computer. [1]Among the characteristics of modern embedded systems are: User Interfaces, Simple Systems which Stem from Limited Functionality and CPU Platforms with Microprocessors or Microcontrollers

User Interfaces

Originally, an embedded system had no user interface-information and programs were already incorporated into the system (e.g., the guidance system for an Intercontinental Ballistic Missile or ICBM) and there was no need for human interaction or intervention except to install the device and test it.

Many modern embedded systems however, have full-scale user interfaces although these are only inputs for data but are not supposed to provide additional functionality for the system, e.g. QWERTY keyboards for PDAs used to enter names, addresses, phone numbers and notes and even full sized documents. The moment PDAs achieve full desktop computer functionalities, however, they may no longer be considered embedded systems.

Simple Systems which Stem from Limited Functionality

Originally, this referred to basic systems such as switches, small character- or digit-only displays and LEDs intended to show the "health" of the embedded system, but this has also achieved some level of complexity. [2]A cash register or an ATM with touch screen technology is considered an embedded system since it has limited uses, even if the user interface (the touch screen) is a complex system.

CPU Platforms with Microprocessors or Microcontrollers

Again, limited functionality is the key in defining these as embedded systems. In a sense, the BIOS chip is considered an embedded system since it has limited functions, and works automatically (when the computer is booted up). Peripherals like the USB can also be considered as embedded systems.

Words

ballistic	*adj.*	弹道的，弹道学的
boot	*v.*	启动

full-scale	*adj.*	完全的，与原物大小一样的，全面的
handset	*n.*	电话听筒，手机，手持机
intercontinental	*adj.*	大陆间的，洲际的
intervention	*n.*	介入，调停，干涉
standalone	*adj.*	独立的，单独的
stem	*v.*	源自，起源于

Abbreviations

BIOS (Basic Input Output System)　　　　[计]基本输入输出系统
USB (Universal Serial Bus)　　　　　　　Intel 公司开发的通用串行总线架构

Notes

[1] 例句：Among the characteristics of modern embedded systems are: user Interfaces, simple systems which stem from limited functionality and CPU platforms with microprocessors or microcontrollers.

　　分析：本句为倒装句。英语中，由于修辞或是出于平衡句子的原因也可以用倒装句。如：Next to this one is another grand hotel which is beautifully decorated. 这家饭店隔壁还有一家装修华丽的大饭店。

　　译文：现代嵌入式系统的主要特点有：用户界面、基于有限功能的简单系统和带有微处理器或微控制器的 CPU 平台。

[2] 例句：A cash register or an ATM with touch screen technology is considered an embedded system since it has limited uses, even if the user interface (the touch screen) is a complex system.

　　分析：consider 作"认为"解时，后面的 an embedded system 是主语 A cash register or an ATM with touch screen technology 的补语。

　　译文：虽然用户界面(触摸屏)是一个复杂的系统，使用触摸屏技术的收银机或自动柜员机仍被认为是嵌入式系统，因为它的用途有限。

Exercises

Ⅰ. Put "true" or "false" in the brackets for the following statements according to the passage.

1. (　) A game controller goes through a series of steps and actions that require less processing power than a standalone computer.

2. (　) The embedded system is supposed to handle complex tasks.

3. (　) The moment PDAs achieve full desktop computer functionalities, they may be considered embedded systems.

4. (　) Originally, an embedded system has a user interface.

5. (　) Many modern embedded systems have full-scale user interfaces which are only inputs for data but are not supposed to provide additional functionality for the system.

6. (　) The key characteristic of an embedded system is that it is supposed to handle a few

simple tasks.

7. (　) Originally, programs were already incorporated into the embedded system.

8. (　) Limited functionality is the key in defining these as embedded systems.

9. (　) BIOS chip works automatically when the computer is booted up.

10. (　) A cash register or an ATM with touch screen technology has limited uses.

Ⅱ. Fill in the blanks according to the passage.

1. Originally, an embedded system had no user interface-information and programs were already _____ into the system.

2. A cash register or an ATM with touch screen technology is considered an embedded system since it has _____ uses, even if the user interface (the touch screen) is a _____ system.

3. Originally, in an embedded system there was no need for human _____ or _____ except to install the device and test it.

4. _____ like the USB can also be considered as embedded systems.

5. _____ is the key in defining these as embedded systems.

6. The key characteristic of an embedded system is that it is supposed to handle a _____.

7. The moment PDAs achieve full desktop computer functionalities, they may no longer be considered _____.

8. A cash register or an ATM with touch screen technology is considered an embedded system since _____.

9. A game controller goes through a series of steps and actions that require as much processing power as a _____.

10. The BIOS chip is considered _____ since it has limited functions.

Ⅲ. Translate the following words and expressions into Chinese.

1. user interfaces
2. limited functionality
3. Intercontinental Ballistic Missile
4. PDA
5. touch screen
6. simple systems
7. CPU platforms
8. LED
9. peripheral
10. cash register

8.2.1 Reading Material

Introduction of "Programming Embedded Systems in C and C++"

This book introduces embedded systems to C and C++ programmers.

Embedded software is in almost every electronic device designed today. There is software hidden away inside our watches, microwaves, VCRs, cellular telephones, and pagers; the military uses embedded software to guide smart missiles and detect enemy aircraft; communications satellites, space probes, and modern medicine would be nearly impossible without it. Of course, someone has to write all that software, and there are thousands of computer scientists, electrical engineers, and other professionals who actually do.

Chapter 8 Embedded Systems

Each embedded system is unique and highly customized to the application at hand. As a result, embedded systems programming is a widely varying field that can take years to master. However, if you have some programming experience and are familiar with C or C++, you're ready to learn how to write embedded software. The hands-on, no-nonsense style of this book will help you get started by offering practical advice from someone who's been in your shoes and wants to help you learn quickly.

The techniques and code examples presented here are directly applicable to real-world embedded software projects of all sorts. Even if you've done some embedded programming before, you'll still benefit from the topics in this book, which include:

- Testing memory chips quickly and efficiently.
- Writing and erasing Flash memory.
- Verifying nonvolatile memory contents with CRCs.
- Interfacing to on-chip and external peripherals.
- Device driver design and implementation.
- Optimizing embedded software for size and speed.

So whether you're writing your first embedded program, designing the latest generation of hand-held whatchamacallits, or simply managing the people who do, this book is for you.

Words

applicable	*adj.*	可适用的，可应用的
customize	*v.*	[计]定制，用户化
hand-held	*adj.*	手提式的，便携式的
hands-on	*adj.*	实用的，与理论相对的
missile	*n.*	导弹，发射物
no-nonsense	*adj.*	实际的，严肃的
nonvolatile	*adj.*	[计]非易失性的
on-chip		片上，片内
pager	*n.*	寻呼机
peripheral	*adj.*	外围的
	n.	外围设备
smart	*adj.*	巧妙的，聪明的，敏捷的
whatchamacallit you	*n.*	指未命名的或说不出名字的某物，是 what may call it 的变化

Phrases

at hand	在手边，在附近，即将到来
be in one's shoes	处在某人的位置
cellular telephone	移动电话

Abbreviations

CRC (Cyclic Redundancy Check)　　　　　　[计]循环冗余码校验
VCR (Video Casstte Recorder)　　　　　　　盒式磁带录象机

8.2.2　正文参考译文

嵌入式系统的主要特点

虽然参与处理或完成这项任务的步骤可能会和其他计算机程序同样复杂，但人们常认为嵌入式系统的一大特点是用来处理一些简单的任务。例如，游戏控制器，可以说有简单的任务：加载游戏，让玩家通过手持装置输入命令来控制游戏。然而，游戏控制器(特别为X-箱或PS3制造的新游戏)的工作是通过了一系列的步骤和操作实现的；执行这些步骤和操作需要与独立的计算机同样大的处理能力。现代嵌入式系统的主要特点有：用户界面、基于有限功能的简单系统、带有微处理器或微控制器的CPU平台。

用户界面

最初，一个嵌入式系统没有用户界面——信息和程序已经纳入系统(例如，洲际弹道导弹或ICBM制导系统)。因此除非安装和测试装置，一般不需要人们干预或人机交互。

但是，许多现代的嵌入式系统有与其他计算机相同的用户接口，虽然这些只是用来进行数据输入，不会为系统提供额外功能。例如：PDA(个人数字助理)的QWERTY键盘用来输入姓名、地址、电话号码和备忘录，甚至是完整的文件。目前PDA已可实现全部台式电脑的功能，所以它们不再被视为嵌入式系统。

基于有限功能的简单系统

最初，嵌入式系统指基本系统，如交换机、小的字符或数字显示器和用于显示嵌入式系统的"健康"的发光二极管，但是这也变得复杂了。虽然用户界面(触摸屏)是一个复杂的系统，使用触摸屏技术的收银机或自动柜员机仍被认为是嵌入式系统，因为它的用途有限。

带有微处理器或微控制器的CPU平台

特定的功能是判定嵌入式系统的关键。从某种意义上说，BIOS芯片被认为是嵌入式系统，因为它有特定的功能并自动运行(当计算机启动时)。同样，外围设备(如USB接口)也可以被看作是嵌入式系统。

8.2.3　阅读材料参考译文

《C、C++嵌入式系统程序设计》介绍

该书向C和C++程序员介绍嵌入式系统的相关知识。

几乎所有电子设备的设计中都有嵌入式软件设计。在我们的手表、微波炉、盒式磁带录像机、移动电话、寻呼机中都隐藏有嵌入式软件，军队应用嵌入式软件进行智能导弹制导，监测敌方飞行器；通信卫星、空间探测、现代医学等都离不开嵌入式软件。这样就需要有人来编写这些软件，现在有很多计算机科学家、电气工程师以及其他专业人士正在从事这项工作。

每个嵌入式系统都是唯一的并且要根据应用要求进行定制。因此，嵌入式系统编程是一个变化范围很宽的领域，需要程序员花好几年的时间掌握。然而，如果你具有编程经验并且熟悉 C 或者 C++，就可以学习开发嵌入式软件。该书可操作性强，风格实用，可以帮助你开始学习，并且从实用的角度出发提供实践建议以帮助你快速学习。

此处给出的技术和代码实例可以直接应用于各种真实的嵌入式软件项目。以前做过一些嵌入式编程的人，仍然可以从该书的内容中获益。该书包括如下主题：
- 快速有效地测试存储芯片的方法。
- 写入或删除闪存的方法。
- 利用循环冗余码校验非易失性存储器内容的方法。
- 片上外设与外部外设的接口。
- 设计与实现设备驱动器的方法。
- 根据体积和速度优化嵌入式软件的方法。

所以，无论你是初次编写嵌入式程序，还是设计最新一代的手持式设备，或者仅仅是管理嵌入式的设计者，这本书都适合你。

8.3 System-level Requirements

In order to be competitive in the marketplace, embedded systems require that the designers take into account the entire system when making design decisions.

1. End-product utility

The utility of the end product is the goal when designing an embedded system, not the capability of the embedded computer itself. [1]Embedded products are typically sold on the basis of capabilities, features, and system cost rather than which CPU is used in them or cost/performance of that CPU. One way of looking at an embedded system is that the mechanisms and their associated I/O are largely defined by the application. Then, software is used to coordinate the mechanisms and define their functionality, often at the level of control system equations or finite state machines. Finally, computer hardware is made available as infrastructure to execute the software and interface it to the external world. [2]While this may not be an exciting way for a hardware engineer to look at things, it does emphasize that the total functionality delivered by the system is what is paramount.

Design challenge:
- Software- and I/O-driven hardware synthesis (as opposed to hardware-driven software compilation/synthesis).

2. System safety & reliability

It is the safety and reliability of the total embedded system that really matters. The Distributed system example is mission critical, but does not employ computer redundancy. Instead, mechanical safety backups are activated when the computer system loses control in order to safely shut down system operation. A bigger and more difficult issue at the system level is software safety and reliability. While software doesn't normally "break" in the sense of hardware, it may be so complex that a set of unexpected circumstances can cause software failures leading to unsafe situations. This is a difficult problem that will take many years to address, and may not be properly appreciated by non-computer engineers and managers involved in system design decisions.

Design challenges:

- Reliable software.
- Cheap, available systems using unreliable components.
- Electronic vs. non-electronic design tradeoffs.

3. Controlling physical systems

The usual reason for embedding a computer is to interact with the environment, often by monitoring and controlling external machinery. In order to do this, analog inputs and outputs must be transformed to and from digital signal levels. Additionally, significant current loads may need to be switched in order to operate motors, light fixtures, and other actuators. All these requirements can lead to a large computer circuit board dominated by non-digital components.

In some systems "smart" sensors and actuators (that contain their own analog interfaces, power switches, and small CPUS) may be used to off-load interface hardware from the central embedded computer. This brings the additional advantage of reducing the amount of system wiring and number of connector contacts by employing an embedded network rather than a bundle of analog wires. However, this change brings with it an additional computer design problem of partitioning the computations among distributed computers in the face of an inexpensive network with modest bandwidth capabilities.

Design challenge:

- Distributed system tradeoffs among analog, power, mechanical, network, and digital hardware plus software.

4. Power management

[3]A less pervasive system-level issue, but one that is still common, is a need for power management to either minimize heat production or conserve battery power. While the push to laptop computing has produced "low-power" variants of popular CPUs, significantly lower power is needed in order to run from inexpensive batteries for 30 days in some applications, and up to 5 years in others.

Design challenge:

- Ultra-low power design for long-term battery operation.

Chapter 8 Embedded Systems

Words

activated	adj.	有活性的，激活的
actuator	n.	激励者，执行器
analog	n.	类似物，相似体，模拟(量、装置、设备)
appreciate	vt.	赏识，鉴赏，感激
	vi.	增值，涨价
competitive	a.	竞争的，比赛的
coordinate	vt.	调整，整理
deliver	vt.	递送，陈述，释放，发表(一篇演说等)
dominate	v.	支配，占优势
equation	n.	方程(式)，等式
finite	a.	有限的
fixture	n.	固定设备，预定日期，[机]装置器
minimize	v.	将……减到最少，[计]最小化
mission	n.	使命，任务
modest	adj.	谦虚的，谦让的，适度的
off-load	vt.	卸载，卸下
	adj.	卸载的，卸下的
paramount	adj.	极为重要的
partition	v.	区分，隔开，分割
pervasive	adj.	普遍深入的
redundancy	n.	冗余
sensor	n.	传感器，探测器
synthesis	n.	综合，合成
tradeoff	n.	(公平)交易，折中，权衡
variant	adj.	不同的
	n.	变量

Phrases

in the face of	不顾，面对，在……前面
shut down	[机](使)机器等关闭

Notes

[1] 例句：Embedded products are typically sold on the basis of capabilities, features, and system cost rather than which CPU is used in them or cost/performance of that CPU.

分析：句中的 rather than，表示客观事实，意为"是……而不是……；与其……不如……"。它连接的并列成分可以是名词、代词、形容词、介词(短语)、动名词、分句、不定式、动词等。例如：

We will have the meeting in the classroom rather than in the great hall. 我们将在教室里开会，而不是在大厅里。

I decided to write rather than (to) telephone. 我最后决定写信而不打电话。

译文：人们购买嵌入式产品是因为它们的功能、特点和系统价值，而不是它们所用的 CPU 或这个 CPU 的成本/性能。

[2] 例句：While this may not be an exciting way for a hardware engineer to look at things, it does emphasize that the total functionality delivered by the system is what is paramount.

分析：句中 what is paramount 为 what 引导的名词性从句，在句中作表语。例如：

The president's meaning is just what we want to express. 主席所表达的意思正是我们所要说的。

译文：虽然硬件工程师早就这样看待事物，但它强调并指出系统提供的总体功能是至关重要的。

[3] 例句：A less pervasive system-level issue, but one that is still common, is a need for power management to either minimize heat production or conserve battery power.

分析：句中 but one that is still common，作 A less pervasive system-level issue 的同位语。

译文：一个不太普遍但仍常见的系统级问题是：电源管理需要最大限度地减少热量的产生，并节约电池电能。

Exercises

Ⅰ. Put "true" or "false" in the brackets for the following statements according to the passage.

1. ( ) Which CPU is used in embedded products or the cost/performance of that CPU determines the sale of embedded products.

2. ( ) The safety and reliability of the total embedded system is of real importance.

3. ( ) Software normally "breaking" in the sense of hardware is a difficult problem that will take many years to address.

4. ( ) Employing an embedded network will reduce the amount of system wiring and number of connector contacts.

5. ( ) Even hardware engineers emphasize the great importance of the total functionality delivered by the embedded system.

6. ( ) Designers take into account the entire system when making design decisions.

7. ( ) The utility of the end product is the goal when designing an embedded system, not the capability of the embedded computer itself.

8. ( ) Embedded products are typically sold on the performance of the CPU.

9. ( ) There is a need for power management to either minimize heat production or conserve battery power.

10. ( ) A difficult issue at the system level is software safety and reliability.

Chapter 8 Embedded Systems

II. Fill in the blanks according to the passage.

1. Software is used to _____ the mechanisms and define their functionality, often at the level of control system equations or finite state machines.

2. A bigger and more difficult issue at the system level is software_____ and _____.

3. The Distributed system example is mission _____, but does not employ computer _____.

4. In order to embed a computer, analog inputs and outputs must be transformed to and from _____ signal levels.

5. A less pervasive system-level but still common issue is a need for power management to either _____ heat production or _____ battery power.

6. The usual reason for embedding a computer is _____.

7. Designers take into account the _____ when making design decisions.

8. Embedded products are typically sold on the basis of _____, features, and _____.

9. It is the _____ of the total embedded system that really matters.

10. Mechanical safety backups are activated when the computer system loses control in order to _____ system operation.

III. Translate the following words and expressions into Chinese.

1. end-product utility
2. mechanical safety backups
3. "smart" sensors
4. coordinate
5. take into account
6. computer redundancy
7. analog inputs and outputs
8. computer circuit board
9. infrastructure
10. off-load

8.3.1 Reading Material

Embedded System Design Issues

Cost and Performance

The embedded systems are developed to perform specific tasks but at lower cost than general-purpose computers. Lowering the cost affects the speed of embedded system. Mostly the functions associated with embedded systems do not require much speed and most often speed issue doesn't matter. Simplifying the hardware allows cost reduction. Using cheaper but slower processors and interfaces such as synchronous serial interfaces can do this.

Speed and Storage

Some tasks may require high speed but not much storage requirements. In such cases small memories can also do the trick. The embedded systems are developed for high volumes. Here the cost becomes a big issue. The selection of memories, chips, controls and integration of CPU is also a concern in cost reduction. The tasks involving small volumes can be achieved by simplifying general purpose PCs. Sometimes general purpose PCs can be used as embedded systems. As an example, rack mount general-purpose computers are used to control devices like printers or drill press. The selection of an embedded system depends upon the requirement specifications.

Specifications and User Constraints for Embedded Systems

Specifications define what task is to be achieved. The specifications contain user constraints. The constraints help the designer to select appropriate hardware and software setup to develop an embedded system. These constraints must be met on real-time. Overall an embedded system performs the specific task according to the real-time constraints using special purpose hardware and embedded software.

Embedded Software/firmware

The hardware requires software to work. The software in embedded systems is known as firmware. The firmwares are developed for embedded systems that do not have disk drives. They are burned on ROMs or flash memories. It is a good practice to avoid the use of mechanical moving parts in embedded systems designing. The embedded systems are designed to perform tasks for years without any errors. Solid-state parts like flash memories are more reliable than those of moving ones such as buttons, switches and disk drives. Moreover, to avoid errors the firmware is tested with much care. For this purpose the coders use simulators. A simulator allows compiling, assembling, running and debugging the code.

Embedded CPU architectures

Atmel AVR, 8051, X86, ARM, MIPS, PIC, PowerPC, Coldfire, H8, SH, V850, FR-V and M32R are most widely used CPU architectures for embedded systems. Each CPU is associated with its own instruction set for programming. One of the most hot embedded system design technique is System On Chip (SOC). System on chip is an application specific Integrated Circuit (ASIC). One of the CPUs mentioned above is used in the IC.

Words

architecture	n.	架构，体系结构
firmware	n.	[计]固件(软件硬件相结合)
implementation	n.	安装，实行，履行
rack	n.	架，破坏
	vt.	放在架上，榨取
specification	n.	规格，详述，详细说明书
synchronous	adj.	同时的，同步的
trick	n.	诡计，骗局，恶作剧，窍门
	vt.	欺骗，哄骗

Phrases

drill press   钻床

Abbreviations

ASIC (Application Specific Integrated Circuit)　　[电]特定用途集成电路
SOC (System On Chip)　　片上系统

8.3.2 正文参考译文

系统级的要求

为了在市场中保持竞争力，嵌入式系统要求设计者在设计时要考虑到整个系统。

1. 最终产品的效用

在设计嵌入式系统时，目的应是提高最终产品的效用而不是嵌入式计算机本身的能力。人们购买嵌入式产品首先是因为它们具有的功能、特点和系统价值，而不是它们所用的 CPU 或这个 CPU 的成本/性能，因为我们可以认为嵌入式系统的机制及其相关的 I/O 主要由应用限定。其次，通常在控制系统方程或有限状态机的层次上，用软件协调这些机制并确定其功能。最后，电脑硬件作为基础设施来执行软件的功能，实现软件与外部世界的连接。虽然硬件工程师早就是这样看待事物的，但它强调并指出系统提供的总体功能是至关重要的。

设计的挑战：

- 软件和 I/O 驱动的硬件综合(而不是硬件驱动的软件汇编/综合)。

2. 系统安全性和可靠性

整个嵌入式系统的安全性和可靠性确实非常重要。分布式系统执行的是关键任务，但不是采用计算机冗余。相反，当计算机系统失去控制时，为了安全关闭系统操作，会激活机械安全备份。对于系统来说更大、更棘手的问题是软件的安全性和可靠性。虽然软件通常不会有硬件意义上的"断裂损坏"，但由于它太复杂，有时一系列意外情况可能会导致软件故障并导致出现不安全。这是一个需要花费许多年才能解决的难题，参与系统设计决策的非计算机工程师和管理人员可能并没有认识到这一问题。

设计的挑战：

- 可靠的软件。
- 现有的、价格低廉的使用不可靠部件的系统。
- 电子与非电子设计的权衡。

3. 控制物理系统

使用嵌入计算机的通常原因是通过监测和控制外部机械与环境互相作用。为了做到这一点，模拟输入要转换为数字信号输入，而数字信号输出要转换为模拟输出。此外，为了操作电机、灯具和其他驱动器，可能需要放大驱动电流。满足这些要求需要设计一个以非数字部件为主的大型计算机电路板。

在一些系统中，"智能"传感器和驱动器(包含自己的模拟接口、电源开关和小 CPU)可用于从中心嵌入式计算机中卸下接口硬件。这带来更多的优势：通过采用嵌入式网络而不是一堆模拟信号线，减少了系统的布线和连接器接点。然而，这种变化也带来了计算机设计的其他问题，即如何在分布式计算机之间分配计算具有低带宽能力的廉价网络。

设计的挑战：

- 分布式系统在模拟、电力、机械、网络和数字硬件加软件之间进行权衡。

4. 电源管理

一个不太普遍但仍常见的系统级问题是：电源管理需要最大限度地减少热量的产生，并节约电池电能。虽然随着便携计算的推出产生了"低功耗"的流行处理器的"变体"，但仍需要特低功耗以实现由电池供电，包括在某些应用中可使用 30 天的价格又不高的电池，以及在其他应用中使用 5 年的电池。

设计的挑战：

- 为长期用电池供电设计超低功耗设备。

8.3.3　阅读材料参考译文

嵌入式系统的设计问题

成本和性能

开发嵌入式系统是为了以低于通用计算机的成本执行特定的任务。降低成本通常会影响到嵌入式系统的速度。执行与嵌入式系统有关的功能通常并不需要多快的速度，多数情况下速度问题并不重要。简化硬件可以降低成本，用速度较慢且价格便宜的处理器和接口，如同步串行接口，就可以做到这一点。

速度和存储

一些任务可能需要运行得快些，但不需要太大的存储。在这种情况下，选用小的存储器也可以。嵌入式系统是为大量应用而开发的，这里成本成了一个大问题。降低成本可以考虑选择存储器、芯片、控制器和 CPU 集成。少量任务可以通过简化通用计算机完成。有时通用计算机可作为嵌入式系统使用，例如，机架式通用计算机用于控制设备(如打印机或钻床)。对嵌入式系统的选择主要取决于用户的具体要求(规格)。

嵌入式系统的规格和用户约束

规格限定要完成什么任务，包含用户约束。这些约束帮助设计者选择适当的硬件和软件装备，以开发嵌入式系统。用户约束必须满足实时的条件。总的来说，嵌入式系统根据实时限制，使用专用硬件和嵌入式软件执行具体的任务。

嵌入式软件/固件

硬件需要软件才能进行工作。嵌入式系统中的软件被称为固件。这些固件为没有磁盘驱动器的嵌入式系统而开发，被烧写在 ROM 或闪存中。在嵌入式系统设计中，这是避免使用机械移动部件的好方法。嵌入式系统是为多年无错误地执行任务而设计的。与移动部件(如按钮、开关和磁盘驱动器)相比，闪存之类的固态部件更可靠。此外，为避免出现错误，要非常认真地测试固件，基于此，编码员一般使用模拟器，模拟器可以用来编译、装配、运行和调试代码。

嵌入式 CPU 架构

嵌入式系统最广泛使用的 CPU 架构包括 Atmel AVR、8051、x86、ARM、MIPS、PIC、PowerPC、ColdFire、H8、SH、V850、FR-V 和 M32R。每个 CPU 有它自己的编程指令集。其中最热门的嵌入式系统设计技术是系统级芯片(SoC，又称片上系统)。系统级芯片是一种专用集成电路(ASIC)，上述的 CPU 都可用于这类集成电路。

8.4 Application of Embedded System

Embedded systems span all aspects of modern life and there are many examples of their use. as shown in Figure 8.2 Telecommunications systems employ numerous embedded systems from telephone switches for the network to mobile phones at the end-user. Computer networking uses dedicated routers and network bridges to route data.

Figure 8.2 Examples of Embedded System Products

Consumer electronics include personal digital assistants (PDAs), MP3 players, mobile phones, videogame consoles, digital cameras, DVD players, GPS receivers, and printers. Many household appliances, such as microwave ovens, washing machines and dishwashers, are including embedded systems to provide flexibility, efficiency and features. Advanced HVAC systems use networked thermostats to more accurately and efficiently control temperature that can change by time of day and season. [1]Home automation uses wired- and wireless-networking that can be used to control lights, climate, security, audio/visual, surveillance, etc., all of which use embedded devices for sensing and controlling.

Transportation systems from flight to automobiles increasingly use embedded systems. New airplanes contain advanced avionics such as inertial guidance systems and GPS receivers that also have considerable safety requirements. Various electric motors — brushless DC motors, induction motors and DC motors — are using electric/electronic motor controllers. Automobiles, electric vehicles, and hybrid vehicles are increasingly using embedded systems to maximize efficiency and reduce pollution.

Medical equipment is continuing to advance with more embedded systems for vital signs monitoring, electronic stethoscopes for amplifying sounds, and various medical imaging (PET, SPECT, CT, MRI) for non-invasive internal inspections.

[2]In addition to commonly described embedded systems based on small computers, a new

class of miniature wireless devices called motes are quickly gaining popularity as the field of wireless sensor networking rises. Wireless sensor networking, WSN, makes use of miniaturization made possible by advanced IC design to couple full wireless subsystems to sophisticated sensor, enabling people and companies to measure a myriad of things in the physical world and act on this information through IT monitoring and control systems. These motes are completely self contained, and will typically run off a battery source for many years before the batteries need to be changed or charged.

Words

span	v.	持续，贯穿，跨越
thermostat	n.	恒温器
surveillance	n.	监视，监督
avionics	n.	航空电子学(航空用电子设备，控制系统)
inertial	adj.	不活泼的，惯性的
hybrid	adj.	混合的
mobile	adj.	可移动的，易变的，机动的
mote	n.	尘埃，微粒
myriad	n.	极大数量
ventilate	vt.	使通风，给……装通风设备

Abbreviations

GPS(Global Position System)	全球定位系统
HVAC(heating, ventilating, and air conditioning)	空气调节
PET, SPECT, CT, MRI	几个医学词汇

Notes

[1] 例句：Home automation uses wired- and wireless-networking that can be used to control lights, climate, security, audio/visual, surveillance, etc., all of which use embedded devices for sensing and controlling.

分析：在定语从句中的 all of which 中，of which 作定语从句主语 all 的定语 which 指代 wired- and wireless-networking。在"介词+关系代词"引导的定语从句中，"介词+关系代词"前可用 some, any, none, both, all, neither, most, each, few 等代词或者数词作定语从句的主语。例如：He has a lot of story-books, a few of which I have never read. 他有很多故事书，有几本我还从未读过。

译文：家庭自动化使用能够控制灯光、气候、安全、音频/视频、监视等的有线和无线网络，所有的网络都使用嵌入式设备进行传感和控制。

[2] 例句：In addition to commonly described embedded systems based on small computers, a new class of miniature wireless devices called motes are quickly gaining popularity as the field of wireless sensor networking rises.

分析：句中 as the field of wireless sensor networking rises 为 as 引导的时间状语从句。as 引导的时间状语从句，往往表示主句动作发生的背景或条件，常常翻译成"随着……"之意。例如：As the time went on, the weather got worse. 随着时间的推移，气候更加糟糕。

译文：除了经常描述的基于小电脑的嵌入式系统，一类被称为 motes(微尘)的新微型无线装置正随着无线传感器网络领域的发展而被迅速普及。

Exercises

Ⅰ. Put "true" or "false" in the brackets for the following statements according to the passage.

1. (　) More embedded systems for vital signs monitoring, electronic stethoscopes for amplifying sounds, and various medical imaging (PET, SPECT, CT, MRI) for non-invasive internal inspections lead to the development of medical equipment.

2. (　) All wired- and wireless-networking, used to control lights, climate, security, audio/visual, survellience, use embedded devices for sensing and controlling.

3. (　) Motes are quickly becoming popular as the field of wireless sensor networking rises.

4. (　) Motes refer to the commonly described embedded systems based on small computers.

5. (　) The motes are hardly self contained, and will typically run off a battery source for many years before the batteries need to be changed or charged.

6. (　) Consumer electronics are including embedded systems to provide flexibility, efficiency and features.

7. (　) Home automation uses embedded devices for sensing and controlling.

8. (　) Automobiles are increasingly using embedded systems to maximize efficiency and reduce pollution.

9. (　) Embedded systems just span some aspects of modern life.

10. (　) Various electric motors are using electric/electronic motor controllers.

Ⅱ. Fill in the blanks according to the passage.

1. Telecommunications systems employ numerous embedded systems from telephone switches for _____ to mobile phones at _____.

2. Many household appliances, such as microwave ovens, washing machines and dishwashers, are including embedded systems to provide_____, _____ and _____.

3. Advanced HVAC systems use networked _____ to more accurately and efficiently control temperature that can change by time of day and season.

4. Automobiles, electric vehicles, and hybrid vehicles are increasingly using embedded systems to maximize_____ and reduce _____.

5. Motes, a new class of miniature wireless devices is completely _____.

6. Home automation uses wired- and wireless-networking that can be used to control lights,

climate, security, audio/visual, surveillance, etc., all of which use embedded devices for sensing and controlling.

7. A new class of miniature wireless devices called _____ is quickly gaining popularity.

8. _____ from flight to automobiles increasingly use embedded systems.

9. New airplanes contain _____ such as inertial guidance systems and GPS receivers that also have considerable safety requirements.

10. Various electric motors are using electric/electronic _____.

III. Translate the following words and expressions into Chinese.

1. telephone switches
2. household appliances
3. networked thermostats
4. hybrid vehicles
5. inertial guidance systems
6. Wireless sensor networking(WSN)
7. electric vehicle
8. miniaturization
9. videogame consoles
10. personal digital assistants(PDA)

8.4.1 Reading Material

Embedded System Market

Software

The embedded software market continues to enjoy steady growth. While market expansion is occurring across many industries, the growth in consumer electronics, especially in mobile and wireless, continues to have a significant impact on the embedded software market as a whole. VDC estimates that the market for embedded operating systems, bundled tools, and related services reached 905$ million in 2004 - an increase of 20.9% over the previous year. This is driven by strong growth in consumer electronics, as well as continued increases in spending in the military/aerospace, industrial automation, and telecom/datacom segments.

The estimated Average Annual Growth Rate (AAGR) between 2004 and 2009 are 16% for embedded software.

The BCC figures show the software revenue by region. U.S.A. owns the major portion of the embedded software revenue (48%) as it is shown in Figure 8.3. Europe, Japan and Asia share the remaining revenue almost equally having 274, 264 and 315 $ millions revenue, respectively. An interesting fact of this market research is that although USA, Europe and Japan have an aggregated annual growth rate of 15%, Asia (mainly because of the proliferation of China) has a growth rate of almost 25%. Hence, by 2009 the revenue of Asia is estimated that will be close to the one of the USA and almost double of the revenues of Europe and Japan.

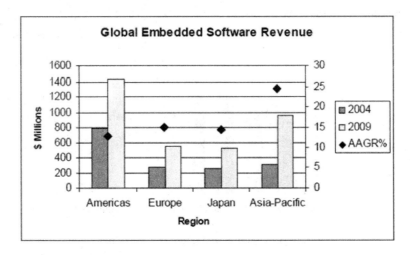

Figure 8.3   Global Embedded Software Revenue by Region, Source: BCC, Inc.

The graph in Figure 8.4 illustrates a strong increase in the value of embedded systems across all sectors. The most prominent examples are telecommunications (with the number of mobile phones rising and added functionalities), logistics, automation.

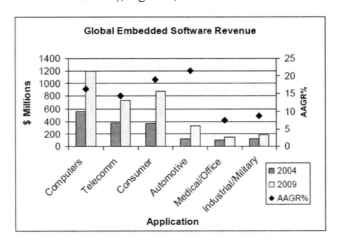

Figure 8.4   Global Embedded Software Revenue by Application, Source: BCC, Inc.

Embedded Operating Systems

Venture Development Corporation estimates worldwide shipments of embedded devices to be over 2 billion in 2004. Talking about current project statistics, this company says that Linux continues to penetrate the embedded market. In a survey done by VDC, nearly 13 percent of respondents indicated that Linux was the primary OS being used for the current project.

But in the LinuxDevices.com's sixth annual Embedded Linux Market Survey , 47 percent of their survey's respondents affirmed to have used Linux in embedded projects and/or products—a growth of about two percentage points over the previous year's results.

While Linux continues to gain popularity, traditional embedded Operating Systems and

Real-Time Operating Systems may be losing marketshare. Trend lines on the chart in Figure 8.5 suggest that by decade's end, actual and planned Linux use will converge, at about 60 percent.

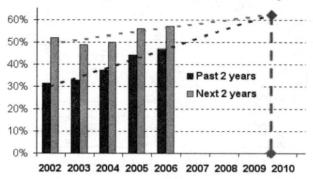

Figure 8.5   Actual and planned LINUX use

The vast majority of embedded Linux developers do not pay anything for their OS, with only 17 percent actually purchasing Linux from a commercial vendor, and the survey shows that it is unlikely to change much over the next two years.

Words

aerospace	*adj.*	航天的，太空的
aggregate	*v.*	聚集，集合，合计
annual	*n.*	年刊，年鉴
	adj.	每年的
converge	*v.*	聚合，集中于一点
enjoy	*vt.*	享受……的乐趣，欣赏，喜爱
impact	*n.*	冲击，影响，效果
military	*adj.*	军事的，军用的
portion	*n.*	部分，一份
proliferation	*n.*	增生(现象)，增生物，扩散
remain	*vi.*	保持，逗留，剩余，残存
respondent	*adj.*	回答的
	n.	回答者
shipment	*n.*	发货，运货，送货，发货量，载货量
steady	*adj.*	稳固的，稳定的，坚定不移的
survey	*n.*	测量，调查，俯瞰
	vt.	调查(收入，民意等)
venture	*n.*	冒险，投机，风险

Abbreviations

DATACOM (Data Communication)	数据通信
VDC (Venture Development Corporation)	文中表示一家公司的名字

8.4.2 正文参考译文

嵌入式系统的应用

嵌入式系统已应用于现代生活的各个领域,生活中使用嵌入式系统的例子很多如图 8.2 所示。从网络电话交换机到最终用户的移动电话,通信系统都使用了多个嵌入式系统。计算机网络使用专用路由器和网桥进行数据发送。

消费类电子产品包括个人数字助理(PDA)、MP3 播放器、移动电话、游戏机、数码相机、DVD 播放器、全球定位系统接收器和打印机。许多家用电器,如微波炉、洗衣机和洗碗机,采用嵌入式系统,保证了灵活性,提高了其效率和功能。先进的空调系统使用网络自动调温器,以更准确、更有效地使温度随时间和季节而变化。家庭自动化使用能够控制灯光、气候、安全、音频/视频、监视等的有线和无线网络,所有这些网络都是使用嵌入式设备进行传感和控制。

从飞机到汽车,运输行业越来越多地使用嵌入式系统。新飞机载有先进的航空电子设备,如惯性制导系统和全球定位系统接收器,这些设备都有相当高的安全要求。各种电动机——无刷直流电动机、感应电动机和直流电动机——使用的是电子/电气马达控制器。汽车、电动车和混合动力车越来越多地使用嵌入式系统,以最大限度地提高效率,减少污染。

医学上更多地使用能监测生命体征的嵌入式系统,放大声音的电子听诊器,用于非侵入性内部检查的各种医疗成像技术(PET、SPECT、CT、MRI),医疗设备也在不断改进。

除了经常描述的基于小电脑的嵌入式系统,一类被称为 motes(微尘)的新微型无线装置正随着无线传感器网络领域的发展而迅速普及。无线传感器网络(WSN)利用先进的 IC 设计所带来的小型化,把完整的无线子系统与先进的传感器结合,使人们和公司能在现实世界中进行许多测量,并根据测量所得信息通过监测和控制系统而采取行动。这些 motes 完全自我控制,且耗能小,更换一次电池或充电一次通常可以使用很多年。

8.4.3 阅读材料参考译文

嵌入式系统市场

软件

嵌入式软件市场继续稳定增长。许多产业的市场在扩大,消费类电子产品,特别是移动和无线领域的增长,继续对嵌入式软件市场整体有重大影响。据 VDC 估计,2004 年嵌入式操作系统、捆绑的工具以及相关服务的市场达到 90.5 亿美元,比 2003 年增加了 20.9%。这是由强劲增长的消费类电子产品,以及继续增加支出的军事/航空航天、工业自动化和电信/数据通信共同推动的。

2004 年至 2009 年间嵌入式软件的估计年平均增长率(AAGR)为 16%。

BCC 公司给出的图表显示了不同地区的软件收入。如图 8.3 所示,美国拥有大部分嵌入式软件的收入(48%)。欧洲、日本和亚洲几乎平分其余收入,分别有 27.4 亿、26.4 亿和 31.5 亿美元的收入。这个市场研究的一个有趣事实是,虽然美国、欧洲和日本综合年增长率为 15%,亚洲的增长速度近 25%(主要是因为中国的发展)。因此,2009 年亚洲的收入估计将接近美国,几乎为欧洲和日本的两倍。

全球不同区域的嵌入式软件收入，来源：BCC 公司(图略)

图 8.4 中的图表说明了嵌入式系统在所有部门增值强劲，其中最突出的部门是电信(由于移动电话数量增加和增加新的功能)、物流、自动化。

全球不同用途的嵌入式软件收入，资料来源：BCC 公司(图略)

嵌入式操作系统

据 Venture Development 公司估计，2004 年全世界嵌入式设备的出货量可能超过了 20 亿美元。在谈到目前项目的统计时，这家公司说，Linux 继续深入嵌入式市场。在该公司所做的调查中，将近 13%的受访者表示，Linux 是用于当前项目的主要操作系统。

在 LinuxDevices.com 的第 6 次年度嵌入式 Linux 市场调查中，47%的受访者确认在嵌入式项目和/或产品中使用了 Linux，与前一年的调查结果相比，增加了约 2 个百分点。

在 Linux 继续普及的同时，传统的嵌入式操作系统和实时操作系统可能失去一定的市场份额。图 8.5 的图表中的趋势线表明，到 2010 年年底，实际和计划的 Linux 使用将趋于一致，大约为 60%。(图略)

这是因为大多数嵌入式 Linux 开发商不为操作系统付费，只有 17%的开发商从运营商处购买 Linux，调查显示在今后两年好像不会有很大的变化。

参 考 文 献

1. 李晓桓，计算机专业英语，武汉：武汉理工大学出版社，2004.
2. 司爱侠，张强华，计算机英语教程，北京：人民邮电出版社，2003.
3. 宋德富，司爱侠，计算机专业英语教程，北京：高等教育出版社，2003.
4. Timothy J.O'Leary, Linda I.O'Leary，计算机专业英语，北京：高等教育出版社，1998.
5. 王道生，韩淑菊，李京，计算机专业英语，北京：电子工业出版社，2003.
6. Leonard Kleinrock, History Of The Internet And Its Flexible Future, IEEE Wireless Communications February 2008.
7. Nell Dale, John Lewis, Computer Science Illuminated, SECOND EDITION.

参 考 文 章

1. http://cordis.europa.eu/ist/embedded/software.htm，2009.5.19
2. http://en.wikipedia.org/wiki/Computer_networking_device，2009.5.19
3. http://en.wikipedia.org/wiki/Database_administrator，2009.5.19
4. http://en.wikipedia.org/wiki/Embedded_system，2009.5.19
5. http://en.wikipedia.org/wiki/Flash_memory，2009.5.19
6. http://en.wikipedia.org/wiki/Internet_service_provider，2009.5.19
7. http://en.wikipedia.org/wiki/Windows_Vista，2009.5.19
8. http://ezinearticles.com/?E-Business-and-Its-Advantages&id=280089，2009.5.19
9. http://users.evitech.fi/~jaanah/IntroC/DBeech/3gl_algorithm1.htm ，2009.5.19
10. http://wapedia.mobi/en/Embedded_system#2，2009.5.19
11. http://www.absoluteastronomy.com/topics/Image_file_formats，2009.5.19
12. http://www.computersciencelab.com/ComputerHistory/HistoryPt4.htm，2009.5.19
13. http://www.deansdirectortutorials.com/Lingo/IntroductionToProgramming.pdf，2009.5.19
14. http://www.ece.cmu.edu/~koopman/iccd96/iccd96.html，2009.5.19
15. http://www.embedsystems.com/，2009.5.19
16. http://www.manythings.org/voa/scripts/2008-06/2008-06-25-voa2.html，2009.5.19
17. http://kb.mit.edu/confluence/display/ist/Introduction+to+Debugging ，2009.5.19
18. http://www.tech-faq.com/embedded-systems.shtml，2009.5.19
19. http://www.tna2000.com/docs/techBulletins/026-HowToInstallTCPIP.Htm，2009.5.19
20. http://www.webopedia.com/quick_ref/OSI_Layers.asp，2009.5.19
21. http://www.websearchworkshop.co.uk/google_history.php，2009.5.19
22. http://www.winnershtriangle.com/w/Articles.WritingSoftwareDocumentation.asp，2009.5.19
23. https://launchpad.net/postgresql，2009.5.19

全国高职高专计算机、电子商务系列教材推荐书目

【语言编程与算法类】

序号	书号	书名	作者	定价	出版日期	配套情况
1	978-7-301-13632-4	单片机C语言程序设计教程与实训	张秀国	25	2012	课件
2	978-7-301-15476-2	C语言程序设计(第2版)(2010年度高职高专计算机类专业优秀教材)	刘迎春	32	2011	课件、代码
3	978-7-301-14463-3	C语言程序设计案例教程	徐翠霞	28	2008	课件、代码、答案
4	978-7-301-16878-3	C语言程序设计上机指导与同步训练(第2版)	刘迎春	30	2010	课件、代码
5	978-7-301-17337-4	C语言程序设计经典案例教程	韦良芬	28	2010	课件、代码、答案
6	978-7-301-09598-0	Java程序设计教程与实训	许文宪	23	2010	课件、答案
7	978-7-301-13570-9	Java程序设计案例教程	徐翠霞	33	2008	课件、代码、习题答案
8	978-7-301-13997-4	Java程序设计与应用开发案例教程	汪志达	28	2008	课件、代码、答案
9	978-7-301-10440-8	Visual Basic程序设计教程与实训	康丽军	28	2010	课件、代码、答案
10	978-7-301-15618-6	Visual Basic 2005程序设计案例教程	靳广斌	33	2009	课件、代码、答案
11	978-7-301-17437-1	Visual Basic程序设计案例教程	严学通	27	2010	课件、代码、答案
12	978-7-301-09698-2	Visual C++ 6.0程序设计教程与实训(第2版)	王丰	23	2009	课件、代码、答案
13	978-7-301-15669-8	Visual C++程序设计技能教程与实训——OOP、GUI与Web开发	聂明	36	2009	课件
14	978-7-301-13319-4	C#程序设计基础教程与实训	陈广	36	2012年第7次印刷	课件、代码、视频、答案
15	978-7-301-14672-9	C#面向对象程序设计案例教程	陈向东	28	2011	课件、代码
16	978-7-301-16935-2	C#程序设计项目教程	宋桂岭	26	2010	课件
17	978-7-301-15519-6	软件工程与项目管理案例教程	刘新航	28	2011	课件、答案
18	978-7-301-12409-3	数据结构(C语言版)	夏燕	28	2011	课件、代码、答案
19	978-7-301-14475-6	数据结构(C#语言描述)	陈广	28	2012年第3次印刷	课件、代码、答案
20	978-7-301-14463-3	数据结构案例教程(C语言版)	徐翠霞	28	2009	课件、代码、答案
21	978-7-301-18800-2	Java面向对象项目化教程	张雪松	33	2011	课件、代码、答案
22	978-7-301-18947-4	JSP应用开发项目化教程	王志勃	26	2011	课件、代码、答案
23	978-7-301-19821-6	运用JSP开发Web系统	涂刚	34	2012	课件、代码、答案
24	978-7-301-19890-2	嵌入式C程序设计	冯刚	29	2012	课件、代码、答案
25	978-7-301-19801-8	数据结构及应用	朱珍	28	2012	课件、代码、答案
26	978-7-301-19940-4	C#项目开发教程	徐超	34	2012	课件
27	978-7-301-15232-4	Java基础案例教程	陈文兰	26	2009	课件、代码、答案
28	978-7-301-20542-6	基于项目开发的C#程序设计	李娟	32	2012	课件、代码、答案

【网络技术与硬件及操作系统类】

序号	书号	书名	作者	定价	出版日期	配套情况
1	978-7-301-14084-0	计算机网络安全案例教程	陈昶	30	2008	课件
2	978-7-301-16877-6	网络安全基础教程与实训(第2版)	尹少平	30	2012年第4次印刷	课件、素材、答案
3	978-7-301-13641-6	计算机网络技术案例教程	赵艳玲	28	2008	课件
4	978-7-301-18564-3	计算机网络技术案例教程	宁芳露	35	2011	课件、习题答案
5	978-7-301-10226-8	计算机网络技术基础	杨瑞良	28	2011	课件
6	978-7-301-10290-9	计算机网络技术基础教程与实训	桂海进	28	2010	课件、答案
7	978-7-301-10887-1	计算机网络安全技术	王其良	28	2011	课件、答案
8	978-7-301-12325-6	网络维护与安全技术教程与实训	韩最蛟	32	2010	课件、习题答案
9	978-7-301-09635-2	网络互联及路由器技术教程与实训(第2版)	宁芳露	27	2010	课件、答案
10	978-7-301-15466-3	综合布线技术教程与实训(第2版)	刘省贤	36	2011	课件、习题答案
11	978-7-301-15432-8	计算机组装与维护(第2版)	肖玉朝	26	2009	课件、习题答案
12	978-7-301-14673-6	计算机组装与维护案例教程	谭宁	33	2010	课件、习题答案
13	978-7-301-13320-0	计算机硬件组装和评测及数码产品评测教程	周奇	36	2008	课件
14	978-7-301-12345-4	微型计算机组成原理教程与实训	刘辉珞	22	2010	课件、习题答案
15	978-7-301-16736-6	Linux系统管理与维护(江苏省省级精品课程)	王秀平	29	2010	课件、习题答案
16	978-7-301-10175-9	计算机操作系统原理教程与实训	周峰	22	2010	课件、答案
17	978-7-301-16047-3	Windows服务器维护与管理教程与实训(第2版)	鞠光明	33	2010	课件、答案
18	978-7-301-14476-3	Windows2003维护与管理技能教程	王伟	29	2009	课件、习题答案
19	978-7-301-18472-1	Windows Server 2003服务器配置与管理情境教程	顾红燕	24	2011	课件、习题答案

【网页设计与网站建设类】

序号	书号	书名	作者	定价	出版日期	配套情况
1	978-7-301-15725-1	网页设计与制作案例教程	杨森香	34	2011	课件、素材、答案
2	978-7-301-15086-3	网页设计与制作教程与实训(第2版)	于巧娥	30	2011	课件、素材、答案

序号	书号	书名	作者	定价	出版日期	配套情况
3	978-7-301-13472-0	网页设计案例教程	张兴科	30	2009	课件
4	978-7-301-17091-5	网页设计与制作综合实例教程	姜春莲	38	2010	课件、素材、答案
5	978-7-301-16854-7	Dreamweaver 网页设计与制作案例教程(2010年度高职高专计算机类专业优秀教材)	吴 鹏	41	2012	课件、素材、答案
6	978-7-301-11522-0	ASP .NET 程序设计教程与实训(C#版)	方明清	29	2009	课件、素材、答案
7	978-7-301-13679-9	ASP .NET 动态网页设计案例教程(C#版)	冯 涛	30	2010	课件、素材、答案
8	978-7-301-10226-8	ASP 程序设计教程与实训	吴 鹏	27	2011	课件、素材、答案
9	978-7-301-13571-6	网站色彩与构图案例教程	唐一鹏	40	2008	课件、素材、答案
10	978-7-301-16706-9	网站规划建设与管理维护教程与实训(第2版)	王春红	32	2011	课件、答案
11	978-7-301-17175-2	网站建设与管理案例教程(山东省精品课程)	徐洪祥	28	2010	课件、素材、答案
12	978-7-301-17736-5	.NET 桌面应用程序开发教程	黄 河	30	2010	课件、素材、答案
13	978-7-301-19846-9	ASP .NET Web 应用案例教程	于 洋	26	2012	课件、素材
14	978-7-301-20565-5	ASP.NET 动态网站开发	崔 宁	30	2012	课件、素材、答案
15	978-7-301-20634-8	网页设计与制作基础	徐文平	28	2012	课件、素材、答案
16	978-7-301-20659-1	人机界面设计	张 丽	25	2012	课件、素材、答案

【图形图像与多媒体类】

序号	书号	书名	作者	定价	出版日期	配套情况
1	978-7-301-09592-8	图像处理技术教程与实训(Photoshop 版)	夏 燕	28	2010	课件、素材、答案
2	978-7-301-14670-5	Photoshop CS3 图形图像处理案例教程	洪 光	32	2010	课件、素材、答案
3	978-7-301-12589-2	Flash 8.0 动画设计案例教程	伍福军	29	2009	课件
4	978-7-301-13119-0	Flash CS 3 平面动画案例教程与实训	田启明	36	2008	课件
5	978-7-301-13568-6	Flash CS3 动画制作案例教程	俞 欣	25	2011	课件、素材、答案
6	978-7-301-15368-0	3ds max 三维动画设计技能教程	王艳芳	28	2009	课件
7	978-7-301-18946-7	多媒体技术与应用教程与实训(第2版)	钱 民	33	2012	课件、素材、答案
8	978-7-301-17136-3	Photoshop 案例教程	沈道云	25	2011	课件、素材、视频
9	978-7-301-19304-4	多媒体技术与应用案例教程	刘辉珞	34	2011	课件、素材、答案
10	978-7-301-20685-0	Photoshop CS5 项目教程	高晓黎	36	2012	课件、素材

【数据库类】

序号	书号	书名	作者	定价	出版日期	配套情况
1	978-7-301-10289-3	数据库原理与应用教程(Visual FoxPro 版)	罗 毅	30	2010	课件
2	978-7-301-13321-7	数据库原理及应用 SQL Server 版	武洪萍	30	2010	课件、素材、答案
3	978-7-301-13663-8	数据库原理及应用案例教程(SQL Server 版)	胡锦丽	40	2010	课件、素材、答案
4	978-7-301-16900-1	数据库原理及应用(SQL Server 2008 版)	马桂婷	31	2011	课件、素材、答案
5	978-7-301-15533-2	SQL Server 数据库管理与开发教程与实训(第2版)	杜兆将	32	2010	课件、素材、答案
6	978-7-301-13315-6	SQL Server 2005 数据库基础及应用技术教程与实训	周 奇	34	2011	课件
7	978-7-301-15588-2	SQL Server 2005 数据库原理与应用案例教程	李 军	27	2009	课件
8	978-7-301-16901-8	SQL Server 2005 数据库系统应用开发技能教程	王 伟	28	2010	课件
9	978-7-301-17174-5	SQL Server 数据库实例教程	汤承林	38	2010	课件、习题答案
10	978-7-301-17196-7	SQL Server 数据库基础与应用	贾艳宇	39	2010	课件、习题答案
11	978-7-301-17605-4	SQL Server 2005 应用教程	梁庆枫	25	2010	课件、习题答案

【电子商务类】

序号	书号	书名	作者	定价	出版日期	配套情况
1	978-7-301-10880-2	电子商务网站设计与管理	沈凤池	32	2011	课件
2	978-7-301-12344-7	电子商务物流基础与实务	邓之宏	38	2010	课件、习题答案
3	978-7-301-12474-1	电子商务原理	王 震	34	2008	课件
4	978-7-301-12346-1	电子商务案例教程	龚 民	24	2010	课件、习题答案
5	978-7-301-12320-1	网络营销基础与应用	张冠凤	28	2008	课件、习题答案
6	978-7-301-18604-6	电子商务概论（第2版）	于巧娥	33	2012	课件、习题答案

【专业基础课与应用技术类】

序号	书号	书名	作者	定价	出版日期	配套情况
1	978-7-301-13569-3	新编计算机应用基础案例教程	郭丽春	30	2009	课件、习题答案
2	978-7-301-18511-7	计算机应用基础案例教程(第2版)	孙文力	32	2012 第2次印刷	课件、习题答案
3	978-7-301-16046-2	计算机专业英语教程(第2版)	李 莉	26	2010	课件、答案
4	978-7-301-19803-2	计算机专业英语	徐 娜	30	2012	课件、素材、答案

电子书(PDF 版)、电子课件和相关教学资源下载地址：http://www.pup6.cn，欢迎下载。
联系方式：010-62750667，liyanhong1999@126.com，linzhangbo@126.com，欢迎来电来信。